Paolo Brandimarte · Agostino Villa
(Eds.)

Modeling Manufacturing Systems

From Aggregate Planning to Real-Time Control

With 54 Figures
and 16 Tables

Springer

Professor Paolo Brandimarte
Professor Agostino Villa

Technical University of Torino
Dipartimento di Sistemi di Produzione
ed Economia dell'Azienda
Corso Duca degli Abruzzi 24
I-10129 Torino, Italy

ISBN 978-3-642-08483-6

Cataloging-in-Publication Data applied for
Die Deutsche Bibliothek - CIP-Einheitsaufnahme
Modeling manufacturing systems : from aggregate planning to real time control / ed.: Paolo Brandimarte; Agostino Villa. - Berlin; Heidelberg; New York; Barcelona; Hongkong; London; Mailand; Paris; Singapur; Tokio : Springer, 1999

Hardcover design: Erich Kirchner, Heidelberg

Preface

Four years ago the International Federation of Automatic Control (IFAC) set up a Technical Committee on Manufacturing Modelling, Management and Control. Among the goals of this committee were:

- the development, the comparison, and the classification of formal models, both descriptive and prescriptive, of Computer Integrated Manufacturing Systems;
- the integration among optimization methods, simulation models and knowledge-based procedures;
- the specification of requirements for new models, including discrete-event and continuous representations, to be used in simulating and designing the management strategies for manufacturing plants.

The technical areas of interest included:

- at the system level, models for plant layout design, process planning, production planning and scheduling;
- at the component level, models for the functional description of flexible manufacturing and assembly systems, and for the design of strategies for production activity control, process supervision, and maintenance.

The Technical Committee is going on with the organization of Workshops and International Symposia under IFAC sponsorship.

This volume collects some contributions from members of the IFAC Technical Committee on Manufacturing Modelling, Management and Control. We wish to thanks the members for their valuable contributions and the anonymous reviewers for their cooperation in making this book possible.

The editors.

Contents

0

Modeling Manufacturing Systems: an Introduction

P. Brandimarte[1]
A. Villa[1]

0.1 Introduction and overview of the contributions

Even a quick glimpse at the literature on modeling in manufacturing systems shows the diversity of approaches that have been adopted. Models may be classified along many dimensions [4].

Purpose. Some models are used to evaluate the performance of a complex system for a given set of control parameters or for a given system configuration. This is the case of evaluative models such as simulation of queueing network models. The suitable system configuration must be obtained by repeated runs of the model, which on the other hand can be quite accurate. In generative models some simplification is accepted in order to obtain an "optimal" answer by solving an optimization model directly (either by an exact or by a heuristic procedure).

Hierarchical level. Some models are used for system design, both in physical or in logical terms. At the opposite end of the line, we may find more operational models which should be used in real time. At an intermediate level, we may find "tactical" models whose time horizon are weeks or a few months.

Time representation. Models of physical systems are usually based on a continuous time representation, leading to differential equations. This approach is taken in manufacturing systems when the material flow is represented as a continuous flow; this paves the way to optimal control models. More commonly, time is discretized (e.g. in months or weeks) to come up with suitable plans; in this case the representation is based on difference equations, to which constraints can be added in order to write linear programming models (possibly mixed-integer). At the physical level, most manufacturing systems can be considered as a discrete event systems. Discrete event models are easily analyzed by evaluative models, but a graph based representation (in which the nodes represents activities and arcs their precedence relationships) can be used to build generative models.

[1]Dipartimento di Sistemi di Produzione ed Economia dell'Azienda, Politecnico di Torino, Corso Duca degli Abruzzi 24, 10129 Torino, Italy. e-mail: brandimarte@polito.it

Uncertainty representation. Uncertainty affects manufacturing systems at different levels. From the real time control point of view, there is an uncertainty on system availability due to machine breakdowns or due to uncertain processing times. At a higher level, changing market trends are reflected in demand uncertainty both from a quantitative and a qualitative point of view, requiring some adaptation in the strategies. Some planning models simply ignore uncertainty since the necessary corrections are left as a task for lower hierarchical levels such as real time control. Other modeling approaches explicitely deal with uncertainty captured by probability distributions of relevant variables or by a stochastic process representation. In other cases, due to lack of information, a non-stochastic representation of uncertainty is preferred, based e.g. on fuzzy logic.

Given this variety of modeling and solution approaches, there is obviously little hope to cover all the possibilities exhaustively. The aim of this book is to complement other books in the area, both single-authored (such as [1]) and multi-authored (such as [3]). In this section we put the contributions to this volume into a common perspective; in the next section some further readings are suggested to those wishing more information on different approaches.

Discrete-time models are the starting point of **Chapter 1**, *From the Aggregate Plan to Lot-Sizing in Multi-level Production Planning*, by J.-C. Hennet. This contribution presents generative and deterministic models, partially inspired by classical lot-sizing models. The aim is to link different decision levels (from aggregate planning to scheduling) through a set of related models; the suitability of different types of representation depending on the hierarchical level is discussed.

Classical scheduling problems are covered in **Chapter 2**, *Shop floor scheduling in discrete parts manufacturing*, by G.J. Meester, J.M.J. Schutten, S.L. van de Velde, and W.H.M. Zijm. The main point of the contribution is that a suitable graph based model can be used to tackle different sub-problems effectively within a common framework. The result is an effective predictive scheduling architecture based on generative, deterministic and discrete-event models.

A graph based representation of manufacturing systems is also at the root of **Chapter 3**, *Integrating Layout Design and Material Flow Management in Assembly Systems*, by M. Lucertini, D. Pacciarelli, and A. Pacifici. Here deterministic optimization models are proposed both for higher level problems (such as designing the layout) and more operational problems (flow management).

Predictive scheduling models must be often integrated with reactive scheduling systems in order to adapt the schedule to disruptions. This is the focus of **Chapter 4**, *Reactive Scheduling in Real Time Production Control*, by E. Szelke and L. Monostori. Here a survey of the wide variety of approaches proposed for reactive scheduling is presented.

Simulation models for the design of manufacturing systems have a long tradition. However, proper modeling of complex systems, characterized by strong automation through robots and material handling systems is by no way trivial. The integration of traditional simulation techniques with CAD tools, Petri Nets and Information Technology approaches is needed. This the topic of two complementary chapters, i.e. **Chapter 5**, *Simulation within CAD-Environment*, by P. Kopacek, G. Kronreif, and T. Perme, and **Chapter 6**, *Model of Material Handling and Robotics*, by C.-Y. Huang and S.Y. Nof.

Finally, the two last chapters are related to the uncertainty issues in manufacturing systems design and management. In **Chapter 7**, *A Simultaneous Approach for IMS Design: a Possibility Based Approach*, by G. Perrone and S. Noto La Diega, fuzzy logic is proposed as a tool to represent uncertainty in situations where the lack of information prevents the use of a classical stochastic representation. In **Chapter 8**, *Adaptive Production Control In Modern Industries*, by K.N. McKay and J.A. Buzacott, a broad discussion of production control approaches is presented, starting from a historical analysis and pointing out the need for organizational flexibility.

0.2 For further reading

All modeling approaches must be put into some applicative perspective, from the point of view of the end user of the results, i.e. the manager. Therefore, before going deep into (sometimes overly) technical topics, it is necessary to have a knowledge of the typical Operations Management problems; good introductory texts are e.g. [8] and [15]. A more technical treatment of Operations Management issues (including forecasting and plant location) is given in [12]; more recent surveys of these topics (with emphasis on generative optimization models) can be found in [10].

Evaluative models of the experimental type (i.e. simulation models) and the related statistical issues are dealt with in [14]; a book specifically oriented to manufacturing systems is [6]. Analytical evaluative models are covered in [17], which deals both with queueing networks and Petri nets. Readers wishing a broader view of Petri Nets are referred to [7].

Discrete time models in manufacturing are usually associated with planning and lot-sizing models. An recent overview of such models can be found in [11] or [13], where it is shown how such models can also be applied to scheduling problems, which are more commonly associated to discrete-event models.

In this book, there is no chapter dedicated to continuous-time models. A thorough treatment of stochastic continuous-time models is presented in [9], whereas readers interested in a quick overview are referred to [16]. Stochastic discrete-event models are the subject of [5]; finally, stochastic discrete-time models can be coped with by stochastic linear programming techniques such as those covered in [2].

0.3 References

[1] R.G. Askin, C.R. Standridge. 1993. *Modeling and Analysis of Manufacturing Systems*. Wiley, New York.

[2] J.R. Birge, F. Louveaux. 1997. *Introduction to Stochastic Programming*. Springer-Verlag, Berlin.

[3] P. Brandimarte, A. Villa (eds.). 1995. *Optimization Models and Concepts in Production Management*. Gordon & Breach, Basel.

[4] P. Brandimarte, A. Villa. 1995. *Advanced Models for Manufacturing Systems*

Management. CRC Press, Boca Raton, FL.

[5] J.A. Buzacott, J.G. Shanthikumar. 1993. *Stochastic Models of Manufacturing Systems.* Prentice-Hall, Englewood Cliffs, NJ.

[6] A. Carrie. 1988. *Simulation of Manufacturing Systems.* Wiley, Chichester.

[7] F. DiCesare, G. Harhalakis, J.M. Proth, M. Silva, F.B. Vernadat. *Practice of Petri Nets in Manufacturing.* Chapman & Hall, London.

[8] E.A. Elsayed, T.O. Boucher. 1994. *Analysis and Control of Production Systems (2nd ed.).* Prentice-Hall, Englewood Cliffs, NJ.

[9] S.B. Gershwin. 1994. *Manufacturing Systems Engineering.* Prentice Hall, Englewood Cliffs, NJ.

[10] S.C. Graves, A.H.G. Rinnooy Kan, P.H. Zipkin (eds.). 1993. *Logistics of Production and Inventory.* North Holland, Amsterdam.

[11] K. Haase. 1994. *Lotsizing and Scheduling for Production Planning*, LNEMS 408, Springer-Verlag, Berlin.

[12] A.C. Hax, D. Candea. 1984. *Production and Inventory Managament.* Prentice-Hall, Englewood Cliffs, NJ.

[13] C. Jordan. 1996. *Batching and Scheduling*, LNEMS 437, Springer-Verlag, Berlin.

[14] A.M. Law, W.D. Kelton. 1991. *Simulation Modeling and Analysis (2nd ed.).* McGraw-Hill, New York.

[15] J.O. McClain, L.J. Thomas, J.B. Mazzola. 1992. *Operations Management.* Prentice Hall, Englewood Cliffs, NJ.

[16] A. Sharifnia. 1995. Optimal Production Control Based on Continuous Flow Models. In: P. Brandimarte, A. Villa (eds.). 1995. *Optimization Models and Concepts in Production Management*, 153-185. Gordon and Breach, Basel.

[17] N. Viswanadham, Y. Narahari. 1992. *Performance Modeling of Automated Manufacturing Systems.* Prentice-Hall, Englewood Cliffs, NJ.

1

From the Aggregate Plan to Lot-Sizing in Multi-level Production Planning

Jean-Claude Hennet[1]

1.1 Introduction

The hierarchical approach to production planning is classically used to handle the organization and logistics in complex production systems. It distinguishes several decision levels using different descriptions of the production process. The upper-level planning horizon is relatively long, and, since the number of time intervals (or buckets) of this horizon is restricted by computational limitations, these buckets are long compared with production cycle times. At this time scale, discrete flows in manufacturing plants can be approximated by continuous flows describing production processes in an aggregated and simplified way. Product flows are generally associated with families of products (Bitran, Tirupati 1993 [4]). In the discrete-time framework used in this study, these flows generate, over each time bucket, important amounts of products, which can be approximately represented by continuous variables. The optimal planned outputs are computed on the basis of forecasted demands and aggregate resource constraints. The results obtained at the aggregate planning level then have to be disaggregated in time and detailed by products and items, taking into account some basic constraints originating from operational levels. As the time horizon is chosen smaller, hard constraints and conflicts increase and continuous variables are less appropriate for describing real phenomena (Agnetis et al. [2]). Other models, of the discrete event type, are needed to schedule production, supplies and sales. Heterogeneous models are obviously hard to combine, and much research effort have been devoted to constructing integrated planning and scheduling schemes (see e.g. Gershwin 1987 [9], Dauzère-Péres, Lasserre 1994 [7]).

The main purpose of this study is to present several continuous models which can be logically sequenced to organize and optimize the implementation of the production decisions taken at the upper planning level. It is shown that, for the set of planning stages, model integration is possible through

- time disaggregation, generating a short-term planning problem with small buckets, using the first period results of the longer horizon plan.
- product family and aggregate resource disaggregation

[1]LAAS-CNRS , 7 Ave. Colonel Roche, 31077 Toulouse FRANCE

- decomposition of demands for final products into demands for components through an MRP-type mechanisms, defining multi-stage planning problems.

Models coherence and data consistency can be achieved along these three planning stages. They have to stop at the limit of validity of continuous models. In production planning, the longer horizon problem is called the **Aggregate Planning Problem**. Time and products disaggregation are combined to define the **Detailed Planning Problem** (Erschler et al. 1985 [14], Merce et al. [20]). Product structures and lead-times are then combined to the end-item short-term production plan to generate the Master Production Schedule (MPS) (see e.g. Fogarty, Hoffmann 1983 [8]). The time-horizon of the detailed planning problem is called the operational horizon. The role of **Multi-stage Planning** is then to optimize the production of primary and intermediate products over the operational horizon. The limit of validity of continuous states model has been identified as related to the well-known lot-sizing problem. Different formulations of this problem can be found in the literature. These formulations are associated with different assumptions on setup costs and on the size of time buckets (see Salomon, 1991, [21]).

This study proposes a possible approach to maintain global consistency from planning to production. This consistency is achieved through a sequence of optimization stages linked by appropriate data flows. This sequence of decision stages is summarized on Fig. 1.

At the bottom of the planning hierarchy, the batch quantities to be produced over the operational horizon constitute the jobs to be scheduled. At smaller time scales, discrete-event problems and real-time considerations become crucial : particular machines must be assigned to particular operations, and operations have to be precisely sequenced and scheduled. Job scheduling problems are not analyzed in this study. They are simplified by assuming that production is repetitively scheduled within each operational horizon, to meet the total requirement and production constraints over the operational horizon. Such a simplification is valid if feasible schedules always exist in spite of precedence constraints between tasks, and limited resource capacity. Here, it is assumed that inventories, resources and demands are in a good adequation, so that feasible schedules exist almost all the time. Hence, in order to consistently and efficiently organize production, it becomes sufficient to optimize the number of batches (or equivalently the number of setups) for each product within each operational time horizon.

1.2 The Planning Process

1.2.1 The Aggregate Planning Problem (APP)

Problem Description

The objective of the Aggregate Production Planning Problem is to organize, over a medium-term horizon (typically one to several months), the utilization of the macro resources of the firm (work centers, aggregate resources,...) to meet the predicted demands for the families of products of the firm catalogue.

The variables of this problem are, for time periods $k = 1, ..., T$.

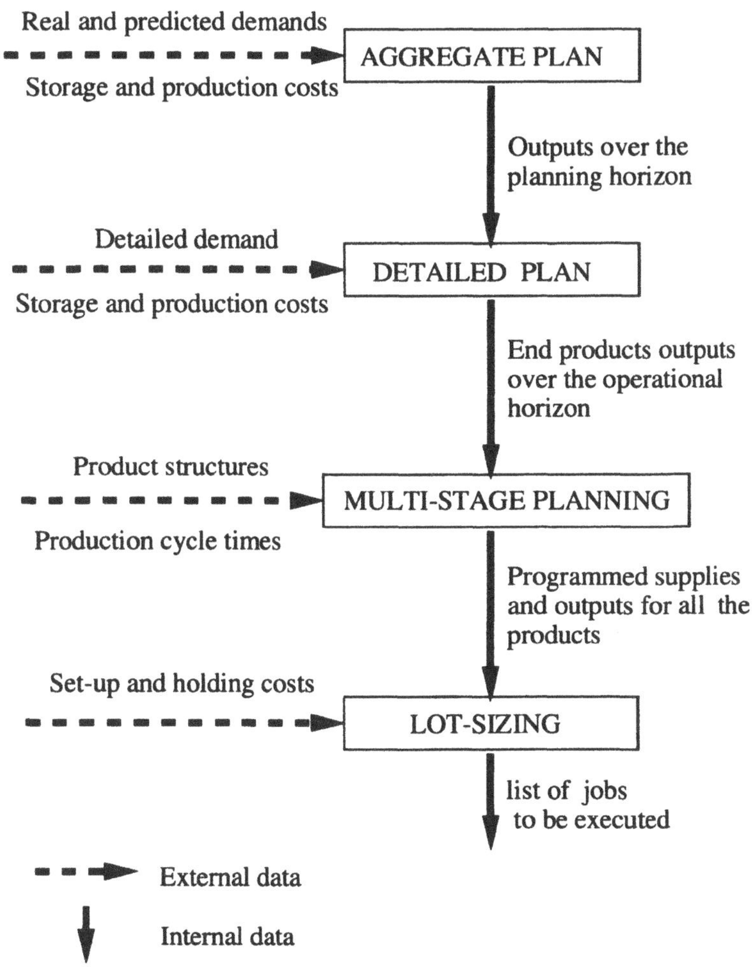

FIGURE 1.1. Data flows along the sequence of decision stages

- the vectors of inventory levels at the beginning of the period, X_k,
- the vectors of production volumes during period k, U_k ; U_{ik} is the reference production level for product-type i at planning period k,
- the vectors of volumes of sales during period k, V_k.

Depending on the type of products manufactured, the typical length of a period could be one day or one weak.

The discrete-time stock evolution equation takes the following form :

$$X_{k+1} = X_k + U_k - V_k. \tag{1.1}$$

Stock and production variables are subject to positivity constraints

$$X_{ik} \geq 0, \quad U_{ik} \geq 0, \ \forall k \in (1, K), \ \forall i \in (1, I).$$

Aggregated production capacity constraints for the R resources and inventory capacity constraints at the P storage spaces can be linearly formulated as follows:

$$\sum_{i=1}^{I} m_i^r U_{ik} \leq \mathcal{M}_k^r \ \forall k \in (1, K), \ \forall r \in (1, R) \tag{1.2}$$

$$\sum_{i=1}^{I} n_i^p X_{ik} \leq \mathcal{N}_k^p \ \forall k \in (1, K), \forall p \in (1, P). \tag{1.3}$$

m_i^r measures the amount of resource r necessary to produce one unit of product i; n_i^p measures the storage requirement at buffer p for storing one unit of product i.

The sequence of vectors V_k representing the volumes of sales is closely related to the vectors of predicted demands D_k. In spite of the possibility of backorders, which is customary in practice, it is generally assumed at the planning level that demand will be satisfied. The equality $V_k = D_k$ is thus imposed, except if together with the other problem constraints, it generates unfeasibility. Predicted demands or resource capacities then have to be modified, or differently distributed in time, to restore feasibility. Under this demand satisfaction constraint, vectors V_k are given, and production volumes are the control variables of the problem.

The Linear Programming Approach

Linear Programming is the most frequently used technique for solving the Aggregate Production Planning Problem. This choice is mainly due to the large size of this problem, to the importance of constraints on vectors X_k, U_k, V_k, and to the linearity of the stock evolution equation (1.1).

A classical linear cost criterion to be minimized over the planning horizon, T, is:

$$\mathcal{J} = \sum_{k=1}^{T} \sum_{i=1}^{I} (c_{ik} U_{ik} + h_{ik} X_{ik}) \tag{1.4}$$

c_{ik} is the unit production cost and h_{ik} is the unit storage cost of product i at period k. For the generality of criterion $\mathcal{J}$, the considered unit costs depend on index k.

They are assumed time-varying. However, in practice, as time variations of unit production and storage costs are hard to predict, it is generally supposed that they are constant over the operational time horizon T.

When using Linear Programming, the Aggregate Production Planning problem is treated as a static problem in which the ordered structure of time does not directly matter. However, in general, only the optimal results X^*_{ik} related to the first time buckets ($k = 1, ..., 5$ for instance, to cover the five days of a working week) are really implemented. The time horizon is used as a receding (or rolling) horizon (see e.g. Bitran and Tirupati 1993 [4]). Then, after the few first buckets, demand and production data are updated, and the planning horizon is shifted to start at the current date. When there are important levels of uncertainty on future demands, raw materials availability, or/and production capacity, the planning problem often has to be run, because Linear Programming solutions are not very robust to variations of data and parameters.

The Predictive Control Approach

Alternative planning techniques such as stochastic control may be better fitted for taking into account random features. Furthermore, stochastic control problems are naturally dynamic : the ordered nature of time is directly accounted for. This property provides the user with more insight into the general trends of the results. On the other hand, control techniques usually present the major drawback of large computational requirements, specially in the presence of many constraints in the problem formulation. They can be preferred to Linear Programming only when the production structure is not too complex.

Among stochastic control techniques, the Model Predictive Control (MPC) approach (Lee et al. 1994 [17]) has shown its practical efficiency in many applications and its ability to integrate constraints in the control design. The choice of the Generalized Predictive Control (GPC) criterion (Clarke et al. 1987 [6]) provides a fair trade-off between demand tracking and production smoothing. In the GPC framework, the sequence of outputs (V_k) is supposed random and unperfectly predicted. The criterion to be minimized over the planning horizon is quadratic :

$$\mathcal{I} = E[\sum_{l=1}^{T}\{(X_{k+l} - X^M_{k+l})'Q_l(X_{k+l} - X^M_{k+l}) + (U_{k+l} - U_{k+l-1})'R_l(U_{k+l} - U_{k+l-1})\}] \tag{1.5}$$

with matrices $Q_l \in \Re^{n*n}$ and $R_l \in \Re^{n*n}$ non negative definite.

Such a criterion contains two terms:

- a penalty term on the differences between the vectors of future inventory levels, X_k and the vectors of reference inventory levels, X^M_k. Weighting matrices Q_l depend on the time index l. The greater importance of first periods results can be easily represented through an approriate choice of these matrices.

- a weighted function of the differences between successive controls. Constraints on production quantities and on production variations can both be represented under this form.

The resulting control law is generally implemented according to an open-loop scheme.

Frequent runs of the program are then needed, specially in the adaptive context. To increase the reactivity of the plan to small demand fluctuations, the predictive control law can also be implemented according to a state-error feedback scheme (Hennet, Barthès 1998 [12]), with the advantage of a better robustness of the production policy. It also allows to integrate state and control constraints (Hennet, 1992 [11]).

1.2.2 THE DETAILED PLANNING LEVEL

In general, product types considered in the Aggregate Production Plan (APP) group several families of products, and each family groups several different items (Bitran, Tirupati 1993 [4]). The aggregation / disaggregation mechanism can then be treated either through linear relations in the case of perfect aggregation (S. Axsater 1981 [3]) or through optimization in the case of approximate aggregation [4], [3].

In the rolling horizon context, criterion (1.5) as well as (1.4) are not *effective* in the sense that the optimal policy obtained from optimization is not fully implemented. The validity of such an approach which truncates an optimal plan can be questioned. As a partial answer to this problem, it has been shown by Lasserre, Bes 1984 [16], that under the assumptions of costs boundedness and finiteness of the control space, there exists a minimal length of the time-horizon above which the first periods optimal results do not change when the optimization horizon is changed.

To illustrate the description of the data transfer mechanism between the aggregate plan, the detailed plan and the multi-stage plan, it can be assumed that only the first period production quantities, $Y(i) = X_{i1}$ for $i = 1, ..., I$ are actually used as reference production levels for the various end-product types $i \in (1, I)$. Production levels, $Y(i)$, for $i \in (1, I)$., generate by disaggregation the final end-products requirements for the detailed planning problem. Under this assumption, the first period of the aggregate plan is also the time-horizon of the multi-stage plan. It is called the *operational horizon.*

1.2.3 THE MULTI-STAGE PLANNING PROBLEM

The detailed production levels obtained by disaggregation determine the reference levels of end-products which should be available by the end of the operational horizon. The objective of the multi-stage planning problem is to program the supply in primary products and the production of intermediate products to achieve the planned production of end-products and to reach the reference levels of safety stocks for all the products. Whether heuristically or optimally solved, the multi-stage planning problem results in what is generally called the Master Production Schedule (MPS) (see e.g. Fogarty, Hoffmann 1983 [8]).

Product structures describe the manufacturing stages and the intermediate products required to obtain the end products. They admit graphical representations, such as *gozinto* graphs (Vazsonyi 1955 [23]) or Petri nets (Vassilaki, Hennet 1986 [22], 1987 [13]). These two graphical tools are represented on Fig.2 to describe the same assembly production structure.

In the Petri net representation, places correspond to stocks, and transitions to assembly and transformation processes. The logics of Petri nets fit well with the logics of production. In particular, an enabled transition corresponds to an operation

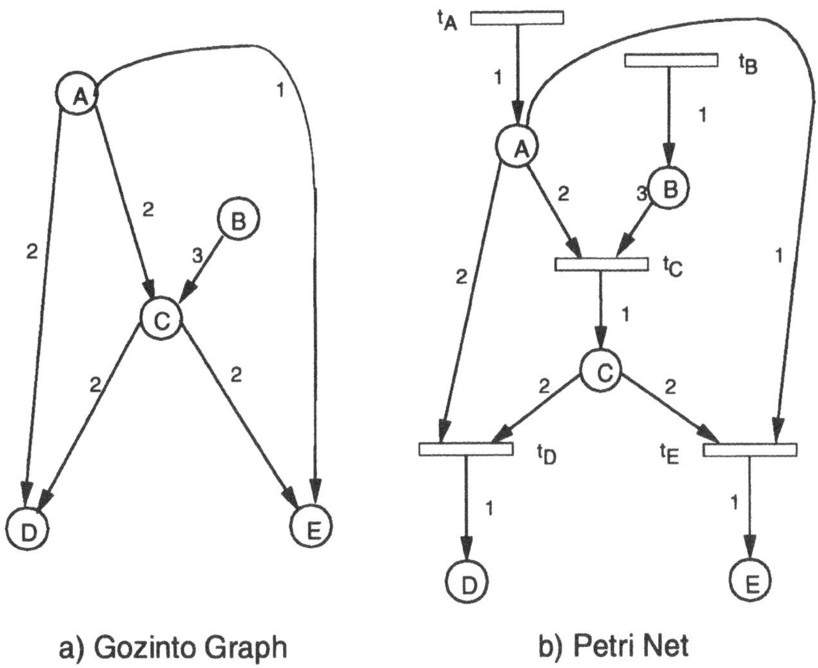

FIGURE 1.2. Two representations of a product structure

which is ready to start, because the input components are in sufficient quantities. High level extensions of Petri nets (see e.g. Jensen, Rozenberg 1991 [15]) also have clear interpretations in manufacturing systems : in particular, different colours may represent different product types, and times associated with transitions may represent operations durations.

Let N denote the total number of products existing in the system, including primary, intermediate and end-products. The combination of end-products reference outputs and safety stocks reference levels can be summarized by a reference vector b, with components $b(j)$, $j = 1, ..., J$, expressing, for all the products $j = 1, ..., J$, the amounts of products which should be available at the end of the current operational horizon. Whenever it is achievable, such a vector b should be obtained as the final state of the multi-stage planning problem. It is called the vector of net outputs for the J products (primary, intermediate and final).

The next planning step is to obtain the gross demand (internal and external) for the J products from the net demand vector b. This step can be performed through an input-output analysis of the system. The input-output approach was originally designed to analyze the interrelations between different sectors of a national economy (Leontief 1951 [18]). Then, it has been successfully applied to multistage production planning problems by several authors, such as Veinott 1969 [1], Grubbström and Molinder 1994 [10].

Production of one unit of product i results from the combination of several production factors and of s_i components in quantities α_{ji}, for $j = 1, ..., s_i$. Production of $y(i)$ units of product i requires $\alpha_{ji}y(i)$ units of component j. Some components are themselves obtained from the combination of other components, and the purpose of the multi-stage planning problem is to determine the necessary quantities of all the products to meet the external requirements $(b(j),\ j = 1, ..., J)$. This static input/output analysis corresponds to the *Bill Of Materials* (BOM). Each column i of matrix A contains the technical coefficients α_{ji} associated with production of 1 unit of component i. For a product structure of the assembly type, the output matrix is the unity matrix, I. The input matrix, A, is square and can be put under an upper-triangular form, with zeros on the main diagonal. Then, vector $y \in \Re_+^J$ of gross outputs for the J products can be computed by the static Leontief equation:

$$y = (I - A)^{-1}b. \tag{1.6}$$

It is not difficult to show that, for an assembly structure, vector y computed by (1.6) is non-negative.

If the production structure is more complex, expression (1.6) may be replaced by the solution of an optimization problem of the following type :

$$\text{Minimize } f(y)$$

$$\text{subject to} \quad (I - A)y \quad = b \tag{1.7}$$

$$y \quad \geq 0 \tag{1.8}$$

Clearly, the BOM analysis is not sufficient to construct the Master Production Schedule, because it does not integrate the timing and sequencing constraints of the multistage problem. However, it provides the sum and substance of what has to be produced within the operational horizon.

The techniques currently used to solve the multi-stage planning problem are Material Requirement Planning (MRP) and Manufacturing Resource Planning (MRP II). Both techniques combine the calculation of required quantities of intermediate products with the treatment of lead-times. However, production and storage are not globally optimized and production capacity constraints are not considered in MRP and heuristically treated in MRP II. An other feature which characterizes MRP-type techniques is that they require a-priori information on lot-sizes for all the products. Such information is external to the multi-stage planning problem, in the sense that the lot-sizing policy is given a-priori. In general, the policy uses fixed lot sizes or is "lot-for-lot".

It is proposed here is to solve the multi-stage planning under the lot-for-lot policy for all the products, but to add a lot-sizing (or lot-streaming) stage after multistage planning. It has been noted by several authors, in particular Dauzère-Péres, Lasserre 1994 [7], that production/inventory costs and cycle times can be notably reduced by splitting production lots into sub-lots. However, two factors limit the lot-straming procedure :

- set-up costs, which are supposedly multiplied by the number of sub-lots,
- lead-times, which are not always proportional to lot-sizes.

The second issue is related to the practical feasibility of the Master Production Schedule. A complete treatment of the combined lot-sizing and scheduling problem has been proposed in Dauzère-Péres, Lasserre 1994 [7]. However, for complex production systems with large numbers of products, a completely integrated technique would not be practical. Rather, it may be assumed that through an appropriate scaling of parameters, the Master Production Schedule can be supposed feasible in the lot-for-lot regime and remain feasible after a limited lot-streaming. Then, after the lot-for-lot resolution of the multi-stage planning problem, a lot-sizing problem can be formulated and solved, to find a good trade-off between reduction of production/inventory costs and increase of set-up costs.

1.3 The Lot-Sizing Problem

1.3.1 PROBLEM DESCRIPTION

At the intermediate level between planning and scheduling, the requirements of the detailed plan can thus be summarized by vector y, given by (1.6) or by the solution of problem (1.7), (1.8). This vector contains the reference production levels of all the components over the "operational horizon", N, which corresponds to one (or several) period of the Aggregate Planning Problem.

A new optimization problem, called the Multiproduct Capacitated Lot Sizing Problem (MCLSP), is then formulated to meet the detailed gross output requirements at minimal cost, while respecting global capacity constraints for each resource (Salomon 1991 [21]).

For output requirements $y(j)$ for items $j = 1, ..., J$, the cost function to be minimized takes the form:

$$\mathcal{C} = \sum_{j=1}^{J} \sum_{k=1}^{N} (c_j u_{jk} + S_j \delta(u_{jk}, u_{j,k-1}) + h_j x_{jk}) \tag{1.9}$$

where

$$\delta(u_{jk}, u_{j,k-1}) = \begin{cases} 1, \text{ if } u_{jk} - u_{jk-1} > 0 \\ 0, \text{ if } u_{jk} - u_{jk-1} \leq 0 \end{cases}, \tag{1.10}$$

x_{j0} and u_{j0}, $\forall\ j \in (1, J)$ are given by the preceding plan in the rolling sequence. The external demand is summarized in vector $y(j)$, which plays the role of a final condition for the state vector x_k :

$$x_{jN} = y(j) \text{ for } j = 1, ..., J. \tag{1.11}$$

This convention of having product requirements only at the end of the operational horizon is natural fort short-term planning.

The three terms of the criterion are production costs, set-up costs and holding costs. The unit cost parameters c_j, S_j, h_j, are supposed time-invariant over this short-term horizon. The non-convexity of this criterion comes from the presence of set-up costs.

The MCLSP slightly differs from the multi-item Capacitated Lotsizing Problem (CLSP) presented in Salomon 1991 [21]. As in the CLSP, production levels for all the items are not necessarily at maximum capacity. Furthermore, a set-up cost S_i is incurred each time a new lot of product i is started within the operational horizon of N periods. But, on the contrary to the CLSP, if production if item j is continued over several successive periods, the associated set-up cost is counted only once. Such an assumption, formalized by definition (1.10), seems quite natural. However, it is seldom considered, partly because it may lead to a rather difficult treatment of boolean variables $\delta(\cdot,\cdot)$.

Under the linear constraints of stock evolution and production capacity, the optimal solution of the MCLSP defines optimal lot sizes and production dates for the various products. This problem is a hard optimization problem with mixed variables. It soon becomes untractable as the complexity of the manufacturing system increases. In the single level case, heuristic techniques such as the ABC method of Maes and Van Wassenhove 1986 [19] have notably increased the dimension of such problems which can be numerically solved. The guideline of the approach proposed here is to try to separate the lot-sizing problem from the lot-scheduling problem. Along this line, a simplified lot-sizing problem will be constructed.

The graph representations of product structures in Fig.2 describe the amounts of components required to produce a given amount of a product. Clearly, such a production cannot be performed if some of the input products are missing. These logical precedence relations are precisely described by the Petri net of Fig.2. If operations durations are taken into account and represented by timed-transitions, then logical precedence constraints generate scheduling constraints. The dynamical evolution of the system can then be described by stock evolution equations with delays (Hennet, Barthès 1998 [12]). However, the criticality of precedence constraints for existence of feasible schedules depends on the production/inventory policy of the firm. They are very important under a "make-to-order" policy, but they become less stringent under "assemble-to-stock" and "make-to-stock" policies. In just-in-time running conditions, components are made available just before the beginning of the assembly task. Such a production rule is not too difficult to obtain in *flat* assembly structures such as the one described in [5] to describe the Toyota goal chasing approach. However, it becomes combinatorially difficult in complex manufacturing systems, with the major drawback of a high sensibility to production and delivery delays.

The complex product structures considered in this study call for some production smoothing tools, which are typically the intermediate stocks. It will thus be assumed that, as long as aggregate capacity constraints are satisfied, which is verified from the APP, feasible schedules generally exist. And if a production sequence is able to provide a given quantity of end-products from an initial state of inventories and to exactly reconstruct these inventories, then, in theory, it can be repeted indefinitely. Therefore, cyclic scheduling guarantees scheduling feasibility if it is built from a feasible and regenerative elementary schedule. As the considered manufacturing systems are large scale and complex, it is natural, at the short-term level of the MCLSP, to consider the following assumptions :

- Assumption 1
  Existence of safety stocks which allow to relax precedence constraints between manufacturing stages (assemble-to-stock assumption). The corresponding production policy is to reconstruct reference stock levels as soon as possible.

- Assumption 2
  Existence of a feasible schedule under the lot-for-lot policy.

- Assumption 3
  Possibility to split the lots while maintaining scheduling feasibility.

The first assumption guarantees that, without oversizing the inventory capacity, the system is permanently driven to a state in which inventory levels are sufficient to sustain assemble-to-stock running conditions for primary and intermediate products, and deliver-to-stock conditions for end products. The second assumptions can be seen as the consequence of the first one and of the existence of sufficient resource capacity. The third one is more technical, but also relies on some type of non-congestion assumption. It may be violated for large numbers of lots if set-up times are not negligible.

Note that, in practice, after the lot-sizing stage, each production lot will be considered as an individual job and scheduling problems may be solved without imposing any cyclic character to the schedule. The three assumpions stated above are simply technical assumptions at the lot-sizing level. They are considered as an approximation of real processing conditions.

Under these assumption, production lead-times do not need to be considered, and the inventory evolution equations for part types take the form:

$$x_{jk+1} = x_{jk} + u_{jk} - \sum_{i=1}^{J} \alpha_{ji} u_{ik} \quad \forall j \in (1, J), \text{ for } k = 1, ..., N-1 \tag{1.12}$$

under boundary conditions:

$$x_{j0} \text{ given ; } \quad x_{jN} = y(j) \ \forall j \in (1, J). \tag{1.13}$$

Stock evolution equations (1.12) are then similar to (1.1), but over shorter time-periods and without any external demand.

Production and storage constraints apply to all the products. They are formally similar to (1.2) and (1.3), using different indices and different bounds:

$$\sum_{j=1}^{J} m_j^r u_{jk} \leq M_r \ \forall k \in (1, N), \ \forall r \in (1, R) \tag{1.14}$$

$$\sum_{j=1}^{J} n_j^p x_{jk} \leq N_p \ \forall k \in (1, N), \forall p \in (1, P). \tag{1.15}$$

$$x_{jk} \geq 0, \quad u_{jk} \geq 0, \ \forall k \in (1, K), \ \forall j \in (1, J)$$

Under the simplifying assumptions of the lot-sizing problem, and possibly using different time origins for the various components, total production requirements can be arbitrarily put at the end of the time horizon, so that the production plan satisfies

$$\sum_{k=1}^{N} u_{jk} = y(j). \tag{1.16}$$

Because of the time-invariant nature of unit costs for this problem, criterion (1.9) can then be re-written:

$$\mathcal{C} = \sum_{j=1}^{J}(c_j y(j) + S_j n_j + h_j \sum_{k=1}^{N} x_{jk}) \tag{1.17}$$

where n_j is the number of production lots of component j over the operational time-horizon. The size of the lots for component j are supposed equal and equal to $\frac{y(j)}{n_j}$. For $y(j)$ given, the first term of the criterion is given and constant. Only the third term explicitely depends on the production schedule.

For each storage space $p \in (1, P)$, the right term of inequality (1.15) is independent of k, and the maximal value of the left term is clearly obtained for $k = N$. Using the final boundary condition in (1.13), constraints (1.15) can thus be equivalently replaced by :

$$\sum_{j=1}^{J} n_j^p y(j) \leq N_p, \quad \forall p \in (1, P). \tag{1.18}$$

Constraints (1.18) impose that lot-for lot production requirements do not exceed the storage capacity at their devoted storage spaces. As a consequence, these constraints guarantee feasibility of the production requirements, from the initial, intermediate and final storage viewpoints, for any number of sub-lots of each product. As constraints (1.18) are purely static, they can be implied by disaggregation of the first period stock capacity constraints (1.3) of the Aggregate Planning Problem.

Production constraints (1.14) can be written:

$$\sum_{j=1}^{J} m_j^r \sum_{l=1}^{k} u_{jl} \leq M_r \; \forall k \in (1, N), \forall r \in (1, R). \tag{1.19}$$

From relations (1.16), a necessary and sufficient condition for constraints (1.19) to hold true at any time period k is the purely static condition:

$$\sum_{j=1}^{J} m_j^r y(j) \leq M_r, \quad \forall r \in (1, R). \tag{1.20}$$

As for production capacity requirements, these constraints can be implied by the first period production capacity constraints (1.2) of the Aggregate Planning Problem. However, in general, constraints (1.20) are necessary, but not sufficient for existence of a feasible schedule. The resources used to process a part may be needed simultaneously, or in sequence. In both cases, because of possible resource sharing by different products, instantaneous production capacity limits may cause the unfeasibility of any schedule. Therefore, conditions (1.20) are part of the MCLSP, but they do not imply satisfaction of scheduling feasibility Assumptions 2 and 3.

If component j is produced at period k, the maximal quantity of j produced during this period (and during any elementary period) is:

$$\mu_{jk} = \mu_j = \min_{r \in (1,R)} \frac{M_r}{m_j^r}. \tag{1.21}$$

The minimal number of periods necessary to produce the required quantity of component j is:

$$N(j) = \frac{y(j)}{\mu_j} \text{ if } \frac{y(j)}{\mu_j} \in \mathcal{N}\ , \quad N(j) = I[\frac{y(j)}{\mu_j}] + 1 \text{ if } \frac{y(j)}{\mu_j} \notin \mathcal{N}, \tag{1.22}$$

where $I(.)$ is the integer part of a real number. A necessary condition for the production requirements to be feasible over the operational time horizon is :

$$N(j) \leq N \ \forall j \in (1, J). \tag{1.23}$$

In particular, the operational horizon should satisfy:

$$N \geq \max_{j \in (1,J)} \frac{y(j)}{\mu_j}. \tag{1.24}$$

The utilization factor of resource r over the operational time-horizon is given by:

$$\rho_r = \frac{\sum_{j=1}^{J} m_j^r y(j)}{N M_r}. \tag{1.25}$$

For the considered production plan, the critical machine, r_c, is defined by: $\rho_{r_c} = \max_{r \in (1,R)} \rho_r$. An other necessary feasibility condition of the set of detailed production requirements, is simply the stability condition:

$$\rho_{r_c} \leq 1. \tag{1.26}$$

Condition (1.26), is generally tighter than condition (1.23). However, in general, it is not a sufficient feasibility condition for the scheduling problem. If several products share several resources, it is not always possible to schedule their processing times so as to satisfy all the resource constraints at each particular time.

Condition (1.26) is a consequence of constraints (1.20). Within the set of assumptions presented in section 3.1, it is satisfied by disaggregation of the first period constraints (1.2) of the APP. In practice, as it is described in the following section, constraint (1.26) can be tested before solving the lot-sizing problem.

1.3.2 A Decomposed Technique for Cost Evaluation

Before solving the lot-sizing problem, it may be useful, as it will be illustrated in the example, to check that production capacity constraints are not violated by the production requirements :

$$\sum_{j=1}^{J} m_j^r (x_{jk+1} - x_{jk}) \leq M_r \ \forall k \in (0, N-1), \ \forall r \in (1, R). \tag{1.27}$$

If production capacity constraints are satisfied, the lot-sizing problem can be rewritten :

$$\text{Minimize } C = \sum_{j=1}^{J} (S_j n_j + h_j \sum_{k=1}^{N} x_{jk})$$

subject to:

$$x_{jN} = y(j) \ \ \forall j \in (1, J) \tag{1.28}$$
$$x_{jk} \geq 0 \forall \ k \in (1, N) \tag{1.29}$$

and subject to additional relations to relate the sequence (x_{jk}) to n_j, which is the number of sub-lots for product j during the operational horizon N.

The lot-sizing problem is now completely decomposed by products. The proposed resolution technique consists of simply using the number of set-ups, n_j, to compute the holding cost for item j, assuming that sequences of production and storage periods are uniformly distributed over a the operational horizon of N elementary periods. In agreement with Assumption 1 in section 3.1, production of each components will be started at maximal rate at the beginning of each production period, so as to maintain net safety stock levels (after internal or external consumption) as close as possible to their nominal values.

Production of component j over the operational time horizon is now simply characterized by the couple $(y(j), n_j)$. Its associated set-up and storage costs can be approximately computed by the following technique:

- The time horizon of N periods is divided into n_j intervals of duration $\frac{N}{n_j}$.
- A production phase at maximal rate is set at the beginning of each interval, as represented on Fig. 2.

  For the ith time interval ($i \in (1, n_j)$), the ith production phase starts at time θ^0_{ij} given by: $\theta^0_{ij} = \frac{(i-1)N}{n_j}$, and the ith production phase ends at time θ^1_{ij} given by: $\theta^1_{ij} = \theta^0_{ij} + \frac{y(j)}{n_j \mu_j}$.

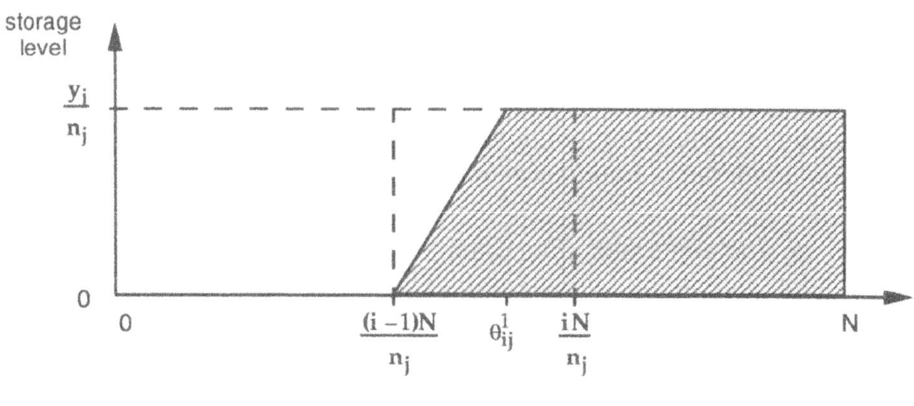

FIGURE 1.3. Definition of average production intervals

The storage cost associated with this production lot over the operational timehorizon is denoted K_{ij}. It is represented by the shaded area of Fig. 2, and computed by:

$$K_{ij} = h_j \frac{y(j)}{n_j}[N - (i-1)\frac{N}{n_j} - \frac{y(j)}{2n_j \mu_j}] \tag{1.30}$$

And the total storage cost associated with production of components j over the operational time-horizon is approximated by K_j, computed by:

$$K_j = \sum_{i=1}^{n_j} K_{ij} = h_j y(j)[N - \frac{(n_j - 1)N}{2n_j} - \frac{y(j)}{2n_j \mu_j}] \tag{1.31}$$

Or, equivalently, $K_j = \frac{h_j y(j)}{2n_j}[(n_j + 1)N - \frac{y(j)}{\mu_j}]$.
Under this simplified evaluation of storage costs, criterion (1.17) takes the form:

$$\mathcal{C} = \sum_{j=1}^{J}[S_j n_j + \frac{h_j y(j)}{2n_j}(N - \frac{y(j)}{\mu_j})] + cst \tag{1.32}$$

where cst is a constant term and n_j are the positive integers to be optimally selected. Positivity of $N - \frac{y(j)}{\mu_j}$ derives from condition (1.24).

As long as production requirements can be exactly met, the optimization problem is totally decomposed by components. The optimal (real) number of lots, n_j^R for item j is obtained from the optimality condition:

$$\left[\frac{\partial \mathcal{C}}{\partial n_j}\right]_{n_j^R} = S_j - \frac{h_j y(j)}{2(n_j^R)^2}(N - \frac{y(j)}{\mu_j}) = 0. \tag{1.33}$$

It is:

$$n_j^R = \sqrt{\frac{h_j y(j)}{2S_j}(N - \frac{y(j)}{\mu_j})} \tag{1.34}$$

Note that, as expected, expression (1.34) is of the same type as the classical Economic Order Quantity (EOQ) formula.

The second derivative of $\mathcal{C}$ with respect to n_j has the sign of $\frac{1}{n_j^3}$. It is always positive for $n_j > 0$. Therefore, criterion (1.32) is convex, and the optimal integer value of n_j, n_j^* is simply given by:

$$n_j^* = I(n_j^R) \text{ or } n_j^* = I(n_j^R) + 1 \tag{1.35}$$

where, as above, $I(.)$ is the integer part of a real number. If $n_j^R \leq 1$, the only possible choice is $n_j^* = 1$. Then, for n_j^R real, the two possible integers values are compared in terms of their contribution to criterion (1.32). Computation of the optimal number of lots for each component type, j, yields the optimal batch quantities, β_j, to be produced: $\beta_j = I[\frac{y(j)}{n_j^R}]$ or $\beta_j = I[\frac{y(j)}{n_j^R}] + 1$.

1.4 Example

This is a simple illustrative example. As the results of the aggregate plan, 3 types of products must be available in final quantities: $b(1) = 7$, $b(2) = 10$, $b(3) = 18$, at the end of the operational horizon of 1 week (48 hours). Product 1 is made from component 4, and product 2 from components 3 and 4. Technological requirements are summarized in the technical matrix:

$$A = \begin{bmatrix} 0 & 0 & 0 & 0 \\ 0 & 0 & 0 & 0 \\ 0 & 1 & 0 & 0 \\ 2 & 1 & 0 & 0 \end{bmatrix}.$$

From (1.6), the vector of total requirements of the 4 components is, with

$$b^T = [7 \quad 10 \quad 18 \quad 0\,],$$

$$y = [I - A]^{-1} b = [7 \quad 10 \quad 28 \quad 24\,]^T.$$

Two machines are devoted to these productions. The resource requirements for producing one unit ot these 4 components are described by vectors:

$$M^1 = [1 \quad 1 \quad 0.5 \;\; 0.5\,]$$

$$M^2 = [0 \quad 2 \quad 0 \quad 1\,],$$

with upper bounds on resources 1 and 2 over each time-period (one hour) : $M_1 = M_2 = 1$. From these requirements, the maximal quantities of items $1, ..., 4$ produced per period are computed using (1.21):

$$\mu_1 = 1 \;,\; \mu_2 = 0.5 \;,\; \mu_3 = 2 \;,\; \mu_4 = 1.$$

The minimal numbers of working periods to produce the items are:

$$N(1) = 7 \;,\; N(2) = 20 \;,\; N(3) = 14 \;\;,\; N(4) = 24$$

Condition (1.24) is satisfied with $N = 48$. The minimal numbers of working periods for machines 1 and 2 are respectively 43 and 44. The corresponding utilization factors are: $\rho_1 \simeq 0.90$, $\rho_2 \simeq 0.92$. Therefore, condition (1.26) is satisfied, but with a relatively small margin.

The holding costs per hour and set-up costs are:

$$h_j \;:\quad 1 \;/\; 2 \;/\; 1 \;/\; 2 \quad ; \quad S_j \;:\quad 10 \;/\; 10 \;/\; 10 \;/\; 10\,.$$

The real lot-splitting parameters obtained by the proposed technique are:

$$n_j^R \;:\; 3.79 \;/\; 5.29 \;/\; 6.90 \;/\; 7.59.$$

Comparison of the integer values around the optimal real ones yields:

$$n_j^* \;:\; 4 \,/\; 5 \,/\; 7 \,/\; 8.$$

In terms of lot sizes, an acceptable trade-off between set-up and storage costs is obtained with the following batch productions:

Product 1 : 3 lots of 2 units, 1 of 1,
Product 2 : 5 lots of 2 units,
Product 3 : 7 lots of 4 units,

Product 4 : 8 lots of 3 units.

The fact of not taking scheduling into account may lead to an overestimations of storage costs, in particular because of the consumption of intermediate components, such as component 4 in the example. But on the other hand, a strict calculation of minimal possible storage costs tends to underestimate real storage costs, specially if it does not take into account safety stocks and possible machine breakdowns. For a real complex production system, the lot-size values given by the proposed technique are nothing but the results of a cost evaluation approach. These results have to be confronted with other views and with technical constraints. Numerical results show that the optimal value of the cost criterion is not too sensitive to lot-size variations around the optimal value. Other criteria can therefore also be considered in practice to improve the lot-size determination.

1.5 Conclusion

In complex production systems, it would be an unrealistic and disputable ambition to achieve complete decisional integration of production activity from the managerial level to shop-floor implementation. However, by an appropriate choice of decision support models and data transfer mechanisms, it is possible to consistently perform product and resource disaggregation and multi-stage production planning. In contrast, the gap between planning and implementation is intrinsic to the different types of variables which have to be considered : quantities of products over given time-periods in planning stages, dates of events and real-time choices in implementation.

In this study, the lot-sizing problem has been analyzed as the ultimate stage of the production planning process. For each component to be produced during a typical operational horizon, the number of lots and their size can be determined in a way which is consistent with the results of the Multistage Planning Problem. This result has been obtained by separating the lot-sizing problem from the lot-scheduling problem. Even if actual scheduling decisions have impacts on inventory and production costs, they are usually too complex and differ too much from planned scheduling decisions to be included in an integrated production planning process.

In the proposed lot-size determination technique, the computation of minimal cost lot-sizes consistent with planned product requirements is performed under an ideal representation of real-time operating sequences as cyclic schedules. But conversely, the number of lots and their size for each component are set to characterize the jobs which remain to be precisely scheduled at the shop floor level.

1.6 References

[1] A.F. Jr. Veinott. Minimum concave-cost solution of Leontief substitution models of multi-facility inventory systems. *Operations Research,* **Vol.17, No 2,** *pp.262-291*, 1969.

[2] A. Agnetis, C. Arbib, and M. Lucertini. The combinatorial approach to flow management in FMS. In *Optimization Models and Concepts in Production*

Management, P. Brandimarte and A. Villa Eds., pages 107–152. Gordon and Breach Publishers, 1995.

[3] S. Axsater. Aggregation of product data for hierarchical production planning. *Operations Research*, 1981.

[4] G.R. Bitran and D. Tirupati. Hierarchical production planning. In *Handbooks in Operations Research and Management Science,* **Vol. 4,** *S.C. Graves, A.H.G. Rinnooy Kan, P.H. Zipkin Eds.*, pages 523–568. North-Holland, 1993.

[5] P. Brandimarte, W. Ukovitch, and A. Villa. Factory level aggregate scheduling: Bridging the gap between optimized scheduling and real time control. In *Optimization Models and Concepts in Production Management, P. Brandimarte and A. Villa Eds.*, pages 187–212. Gordon and Breach Publishers, 1995.

[6] D.W. Clarke, C. Mohtadi, and P.S. Tuffs. Generalized predictive control, parts 1 and 2. *Automatica,* **Vol.23, No.2,** *pp.137-160*, 1987.

[7] S. Dauzère-Péres and J.B. Lasserre. *An integrated approach in production planning and scheduling.* Springer-Verlag, 1994.

[8] D.W. Fogarty and T.R. Hoffmann. *Production and inventory management.* South-Western Publishing Company, 1983.

[9] S.B. Gershwin. A hierarchical framework for manufacturing system scheduling. In *Proceedings of the IEEE Conf. on Decision and Control, Los Angeles*, 1987.

[10] R.W. Grubbström and A. Molinder. Further theoretical considerations on the relationship between mrp, input-output analysis and multi-echelon inventory systems. *Intl. J. Production Economics,* **Vol. 35,** *pp.299-311*, 1994.

[11] J.C. Hennet. A modelling technique for production planning. *Manufacturing Systems,* **Vol.21 , No.2,** *pp.107-112*, 1992.

[12] J.C. Hennet and I. Barthès. Closed-loop planning of multi-level production under resource constraints. In *Proceedings of the IFAC Symposium INCOM'98, Nancy, (France)*, 1998.

[13] J.C. Hennet and M. Vassilaki. Modélisation et gestion de systèmes de production avec stockages. *RAIRO-APII,* **Vol. 21,** *pp.3-16*, 1987.

[14] C. Merce J. Erschler, G. Fontan. Consistency of the disaggregation process in hierarchical planning. *Operations Research,* **Vol. 14, no. 3**, 1985.

[15] K. Jensen and G. Rozenberg Eds. *High-level Petri Nets.* Springer-Verlag, 1991.

[16] J.B. Lasserre and C. Bes. Infinite horizon non stationary stochastic optimal control problem: A planning horizon result. *IEEE Trans. Automatic Control,* **Vol. 29, No.9,** *pp. 836-837*, 1984.

[17] J.H. Lee, M. Morari, and C.E. Garcia. State-space interpretation of model predictive control. *Automatica,* **Vol. 30, No.4,** *pp. 707-717*, 1994.

[18] W.W. Leontief. *Structure of the American Economy 1919-1939.* Oxford University Press, New-York, 1951.

[19] J. Maes and L.N. Van Wassenhove. A simple heuristic for multi-item single level capacitated lotsizing problems. *Operations Research Letters,* **Vol.4, No.6,** *pp.265-273.*, 1986.

[20] C. Merce, G. Hetreux, and G. Fontan. Planification agrégée et planification détaillée. In *Concepts et Outils pour les Systèmes de Production, J.-C. Hennet Ed.*, pages 23–44. Cepadues, 1997.

[21] M. Salomon. *Deterministic Lotsizing Models for Production Planning.* LNEMS **Vol. 35**, Springer Verlag, Berlin, 1991.

[22] M. Vassilaki and J.C. Hennet. Modelling and safety stock analysis in a multi-product industrial process. In *Proceedings of the IMACS-IFAC Symposium, Lille, France, pp.439-442*, 1986.

[23] A. Vazsonyi. The use of mathematics in production and inventory control. *Management Science,* **Vol. 1, No. 1** *pp.70-85*, 1955.

2

Shop Floor Scheduling in Discrete Parts Manufacturing

G.J. Meester[1]
J.M.J. Schutten[1]
S.L. van de Velde[2]
W.H.M. Zijm[1]

ABSTRACT This paper discusses the architecture and the algorithmic framework of an automatic shop floor planning and scheduling system that is currently used in practice. After reviewing some general trends in manufacturing and the role of planning and scheduling systems in particular, it outlines a decomposition framework which forms the basis of the system proposed here. Next, a number of algorithmic enhancements, needed to deal with more complex but realistic machining systems, are discussed, such as multi-resource scheduling techniques, the inclusion of set-up times in scheduling with due dates, techniques for jobs with assembly structures as well as several minor additional features. A brief exposition of the architecture of the system and a discussion of some experiences in practice conclude the paper.

2.1 Introduction

During the last two decades, a revitalization of the role of manufacturing and a renewed awareness of its impact on both national and global economics can be observed. Various circumstances have played a role in raising this renewed interest, such as:

- Changing markets. Starting with the early sixties, the growth in prosperity in Western economies gradually led to a higher variety and customization of products, and to a demand for high quality and short response times (*cf.* Deming [Dem82] and Blackburn [Bla91]). This is often termed the "shift from a seller's to a buyer's market".
- Global competition. In particular for Japan, almost completely lacking natural resources, the energy crises of the seventies marked the starting point in developing production systems, characterized by a high efficiency, an improved quality and a severe reduction of capital tied up in work-in-process inventories (*cf.* Schonberger [Sch82]).

---

[1]Production and Operations Management Group, Dept. of Mechanical Engineering, University of Twente
[2]Department of Decision and Information Sciences, Rotterdam School of Management, Erasmus University

- Manufacturing technologies. Both new materials and automation have had a profound impact on the nature of many productive systems, in design (miniaturization of components), process planning, manufacturing, and distribution. In addition, the evolution in information system technology created new possibilities for manufacturing management.

Influenced by these developments, new manufacturing planning and control paradigms emerged, leading to material coordination systems such as Material Requirements Planning (Orlicky [Orl75]), Manufacturing Resources Planning (Wight [Wig84]) and Just in Time (Schonberger [Sch82]). A more capacity oriented approach, Hierarchical Production Planning, has been advocated for (semi-)process industries by Hax and Meal [HM75], while workload based control systems for job shops were proposed by, *e.g.*, Bertrand and Wortmann [BW81], Bechte [Bec87], Wiendahl [Wie87] and Goldratt [Gol88]. To complete the picture, also more integral views on designing production systems should be mentioned, such as Cellular Manufacturing (Hyer and Wemmerlov [HW89]) and more company-wide approaches like Lean Manufacturing (Womack *et al.* [WJR90]) and Business Process Re-engineering (Hammer and Champy [HC93]).

Unfortunately, for discrete manufacturing systems, in particular for small batch parts manufacturing, these planning systems have severe limitations. MRP I and II are materials coordination systems which lack sufficient capabilities for detailed capacity planning. Basically, these approaches are particularly suitable for relatively stable, medium to large volume manufacturing environments (MRP was initially developed for assembly lines). Hierarchical Production Planning is particularly suitable for planning bottlenecks in process industries but does not support complex Bill of Materials structures occurring in discrete manufacturing. Workload control systems basically are a form of input-output control and as such too global to ensure on-time delivery of complex parts, needing both primary and auxiliary resources simultaneously.

What remains is the application of intelligent planning and scheduling techniques. Arguments to support the implementation of advanced scheduling systems, as opposed to global workload control procedures or simple priority rules, include:

- The increased need to satisfy tight due dates for a large variety of jobs often requires a careful trade-off between clustering jobs, to save set-up times, on the one hand and sequencing jobs according to some urgency measure on the other hand. Classical priority rules are not able to make such a trade-off since they depend on single job, instead of job-group, characteristics, while also machine set-ups are often not taken into account. In addition, isolated operators often do not have the information available to make proper decisions. An integral view on the complete manufacturing cell or shop is needed instead.

- The simultaneous use of several resources introduce a coordination problem typical of automated manufacturing systems. In classical machining environments, machines were often equipped with their own set of tools while also operators were seldomly shared. However, in Flexible Manufacturing Cells, sharing of auxiliary resources (tools, pallets, fixtures) and (one or two) operators is the rule rather than the exception. Such coordination problems cannot be handled by simple priority rule based systems or workload control procedures.

Unfortunately, despite the tremendous developments in deterministic scheduling theory and algorithms (see, *e.g.*, Lawler *et al.* [LLK82], Morton and Pentico [MP93], and Pinedo [Pin95]) application of the theory in practice has been limited for a long time. Among the arguments often used against a more profound role of scheduling in manufacturing are:

- Research in scheduling theory has mainly focussed on the analysis of a standardized collection of scheduling problems, which are hardly relevant in practice. Moreover, the main research topics include defining the borderline between 'easy' and 'intractable' problems and the design and analysis of approximation algorithms with worst-case performance guarantees and optimization algorithms of the branch-and-bound type. Practical scheduling problems, more difficult than the standardized theoretical scheduling problems, are often ignored because they do not lend themselves to a profound and elegant mathematical analysis.
- The relevance of any detailed schedule is limited anyhow due to the occurrence of many unforeseen events on the shop floor such as machine breakdowns, the unavailability of auxiliary equipment such as tools or fixtures, disapproval of procured material, etc. Also, the notion of optimality in deterministic scheduling models does not reflect the often existing data inaccuracy (*e.g.*, a limited predictability of processing times in small batch parts manufacturing).

The first argument now has lost much of its validity in view of the recent advances in heuristic scheduling techniques, based on, *e.g.*, simulated annealing (Van Laarhoven *et al.* [LAL92]), taboo search (Glover [Glo89, Glo90]) or genetic algorithms (Della Croce *et al.* [CTV95]). With respect to the second argument, two responses are in place. First, experience in many practical environments indicate that many so-called disturbances have an organizational, rather than a technical background and could have been prevented by using integrated planning techniques (*e.g.*, for scheduling both primary and auxiliary resources, see Tiemersma [Tie92]). Second, the argument merely illustrates that schedules have to be robust against small disturbances and quickly adaptable in case of large disturbances, but by no means rules out the use of advanced scheduling techniques.

Ironically enough, the need to use plans and schedules in discrete manufacturing facilities seems to be undisputed, as shown by the increased use of both manual and automated planning boards in many factories. But an automated planning board is still nothing more than an information system that can graphically display a schedule constructed by a manual planner, and generate various performance measurement reports of interest to such a planner, such as utilization rate, number of tardy jobs, and average lateness. An automatic planning and scheduling system on the other hand is a *Decision Support System*, consisting of the above described planning board features enriched with a set of intelligent algorithms that allow for an automatic generation of high quality schedules, satisfying preset constraints and objectives (but clearly leaving the possibility to intervene open). As already mentioned above, the increased complexities of modern automated manufacturing systems, make the application of advanced scheduling techniques inescapable.

Fortunately, the architecture of these systems creates new possibilities as well. In particular in highly automated manufacturing shops, masses of data have to be transferred between several stations, *e.g.*, the process planning department, the tool preparation room, the machine shop, and the dispatch area, often by means of a local area network. This provides the ideal infrastructure for automatic planning and scheduling, in particular since automatic monitoring and control becomes much easier.

Having thus motivated the application of intelligent planning and scheduling techniques, an example of an advanced shop floor scheduling system is presented in this paper. We discuss the main features of a system for discrete parts manufacturing that has been developed in collaboration with several companies and is now operating in practice. The focus in this paper is on the algorithmic contributions. In Section 2.2, we outline the basic framework and the decomposition technique that form the heart of the system. The next four sections discuss a number of algorithmic enhancements that have grately increased the practical applicability of the system, among which multi-resource scheduling techniques, the inclusion of set-up times in due date scheduling, convergent job routings, and various additional characteristics. An outline of the system's architecture and some experiences in practice are discussed in Section 2.7. Section 2.8 concludes the paper.

2.2 Basic decomposition approach

In this section, we discuss the classical job shop problem, the disjunctive graph to represent any instance of this problem, and the Shifting Bottleneck (SB) procedure of Adams *et al.* [ABZ88], which is a heuristic to solve the classical job shop problem. The SB procedure decomposes the job shop problem into a series of single-machine scheduling problems. After this, we show that by changing the properties of the disjunctive graph and by changing the machine scheduling algorithms, we can extend the SB procedure to cover many practical features. We emphasize, however, that the decomposition *principle* of the SB procedure is maintained.

The problem of scheduling jobs in machine shops is often modeled as a *classical* job shop problem. This problem is concerned with a shop consisting of m machines $M_1, M_2, \ldots, M_m$ on which a set of n jobs $J_1, J_2, \ldots, J_n$ needs to be processed. Each machine is available for processing from time 0 onwards and can process at most one job at a time. Each job J_j consists of a *chain* of operations $O_{1j}, O_{2j}, \ldots, O_{n_j,j}$, where n_j denotes the number of operations of job J_j. Operation O_{1j} is available from time 0 onwards, whereas operation O_{ij} can only be processed after the completion of operation $O_{i-1,j}$ $(i = 2, \ldots, n_j)$. Operation O_{ij} $(j = 1, 2, \ldots, n; i = 1, 2, \ldots, n_j)$ needs uninterrupted processing on a given machine μ_{ij} during a given non-negative time p_{ij}. The objective usually considered in the literature is to find a schedule that minimizes the *makespan* $C_{\max}$, that is, to find a schedule in which the time to process all jobs is minimal.

Each instance of the job shop problem of minimizing makespan can be represented by a disjunctive graph $G = (V, A, E)$, with V a set of nodes, A a set of arcs, and

E a set of orientable edges; this graph is due to Roy and Sussman [RS64]. For each operation O_{ij}, V contains a node v_{ij} with weight p_{ij}; V also contains two auxiliary nodes: a source s and a sink t, both with weight 0. Arc set A consists of arcs between nodes corresponding to successive operations of a job, arcs from node s to the nodes v_{1j}, and arcs from the nodes $v_{n_j,j}$ to t. E consists of edges between nodes corresponding to operations that need to be processed on the same machine. The weights of all arcs and edges are 0. As an example, Figure 2.1 represents the

J_j	μ_{1j}	μ_{2j}	μ_{3j}	p_{1j}	p_{2j}	p_{3j}
J_1	M_1	M_2	M_3	4	7	6
J_2	M_3	M_1	M_2	3	5	8
J_3	M_2	M_3	M_1	2	6	7

TABLE 2.1. Data for example instance.

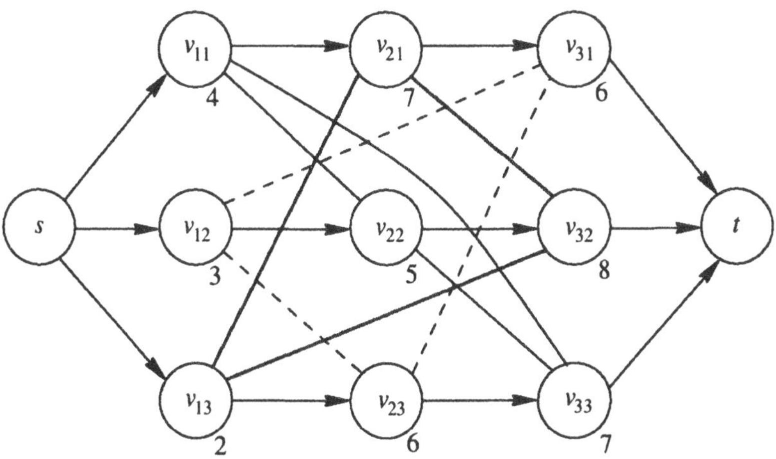

FIGURE 2.1. Graph representing example instance.

instance with three jobs and three machines. Each job consist of three operations. The data of this example can be found in Table 2.1.

A *left-justified* schedule is a schedule in which no operation can start earlier without changing the sequences of jobs on the machines. For the classical job shop problem, there always exists an optimal left-justified schedule. The crux is now that there is a one-to-one relation between a left-justified schedule and a feasible orientation of the edges in E. An orientation of the edges in E is feasible if the resulting graph is acyclic. The makespan of a schedule corresponding to a feasible orientation of the edges in E is equal to the length of a longest path from s to t in the resulting graph. The problem of finding a scheduling with minimum makespan is therefore equal to finding a feasible orientation of the edges in E so that the length of a longest path from s to t in the resulting graph has minimum length.

In practice, jobs have often different release and due dates. We can easily model release and due dates in the graph G. A release date r_j for job J_j $(j = 1, \dots, n)$ is represented in G by giving the arc from s to v_{1j} length r_j. Analogously, a due date

d_j for job J_j is represented in G by giving the arc from $v_{n_j,j}$ to t length $-d_j$. The length of a longest path from s to t is now equal to the *maximum lateness* $L_{\max}(\sigma)$ of the induced schedule σ, with

$$L_{\max}(\sigma) = \max_{j=1,\dots,n} L_j(\sigma) := \max_{j=1,\dots,n} \{C_j(\sigma) - d_j\}$$

and $C_j(\sigma)$ the completion time of J_j in σ.

The classical job shop problem is one of the hardest combinatorial optimization problems. For example, a problem with only 10 jobs and 10 machines, proposed by Fisher and Thompson [FT63], remained unsolved for more than 25 years, in spite of the research effort spent on it. Due to its intractability, several authors developed branch-and-bound algorithms to solve the problem; *cf.* Carlier and Pinson [CP89] and Brucker *et al.* [BJS94]. In other papers, various approximation algorithms have been proposed for the job shop problem, including taboo search, simulated annealing, and genetic algorithms. We refer to Vaessens *et al.* [VAL96] for a computational study of the performance of the most prominent ones. One of them is the Shifting Bottleneck (SB) procedure of Adams *et al.* [ABZ88]. The SB procedure is an intuitive algorithm that decomposes the job shop scheduling problem into a series of single-machine scheduling subproblems. It produces schedules of good quality in reasonable time.

Like most algorithms for the classical job shop problem, the SB procedure heavily relies on the computation of longest paths in the graph G:

- the length of a longest path from node s to node v_{ij} defines the earliest possible starting time of operation O_{ij}, that is, it defines a *release date* r_{ij} for operation O_{ij};
- the length of a longest path from node v_{ij} to node t equals the minimum time the shop needs to process all jobs after the completion of operation O_{ij}, that is, it defines a *run-out time* q_{ij} for operation O_{ij}. By setting $d_{ij} = -q_{ij}$, we get a *due date* for operation O_{ij};
- if all machines are scheduled, then the length of a longest path from s to t equals the maximum lateness of this schedule.

The SB procedure starts by removing all edges from G, labeling all machines as non-bottleneck machines, and computing the longest paths, resulting in release and due dates for each operation. Next, all machines are scheduled separately. We then need to solve m single-machine scheduling problems of minimizing maximum lateness where the operations have release dates. Adams *et al.* use Carlier's [Car82] optimization algorithm to solve these single-machine scheduling problems. The machine with the largest resulting maximum lateness is labeled as a bottleneck machine. The schedule of this machine is fixed by adding the arcs representing the schedule of this machine to G. Now, longest paths are recomputed and the non-bottleneck machines are scheduled subject to the updated release and due dates. The machine that now results in the largest maximum lateness is also labeled as a bottleneck machine. The bottleneck machines are now *rescheduled* in a special bottleneck optimization step. G is changed, such that the machine arcs represent the, possibly changed, schedules of the bottleneck machines. Longest paths are again computed, the non-bottleneck machines are scheduled, and so on. This process continues until all machines have been labeled as bottleneck machines and have been rescheduled in the bottleneck optimization step.

A nice feature of the SB procedure is that it can be extended to deal with various practical features. We already showed how to model release and due dates of jobs. Below, we discuss the extensions to deal with simultaneous resource requirements, set-up times, and convergent and divergent job routings. For a more elaborate discussion of possible extensions of the SB procedure, we refer to Schutten [Sch, Sch96]. We once again stress that we maintain the decomposition principle of the standard SB procedure. For related work, we refer to Ivens and Lambrecht [IL96], Demeulemeester and Herroelen [DH96], and Ovacik and Uzsoy [OU97].

2.3 Multi-resource scheduling

In practice, an operation may need more than one resource simultaneously for its processing. Besides a machine, an operation may need a pallet on which it must be fixed, certain tools, or an operator at the machine. We model this by adding disjunctive edges to G that connect all operations that need the same resource. In Figure 2.2, operations O_{31}, O_{32}, and O_{33} need, besides the machines, the same additional resource. In the SB procedure, we orient those edges in such a way that they

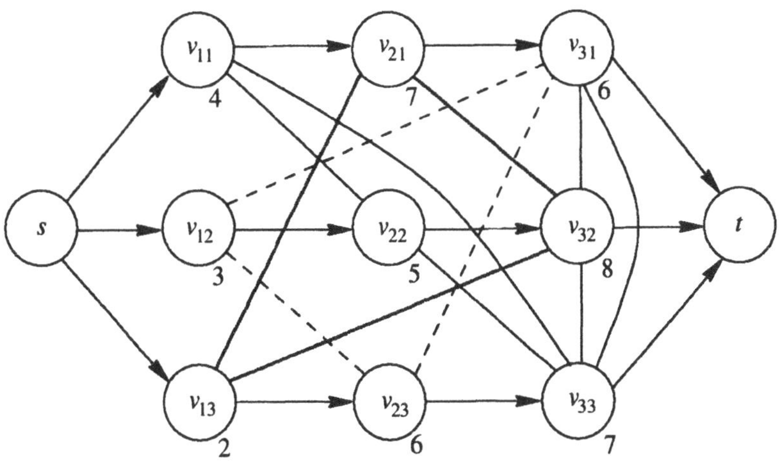

FIGURE 2.2. Graph with multi-resource aspects.

represent the schedules on the additional resources. We distinguish two approaches to deal with multiple resources.

1. *The centralized approach.* In this approach, we treat every resource as a machine that needs to be scheduled. We do not differentiate between machines and other resources. Consequently, every resource becomes a bottleneck machine in the SB procedure. This approach is useful when the number of additional resources is limited. The modeling of this approach affects G only, not the machine scheduling subproblems. The interaction of the different resources is

handled by the SB procedure; this is why we call this approach the centralized approach.

2. *The decentralized approach.* If the number of additional resources is large, then the centralized approach may be time-consuming, since we need to schedule each resource separately. Sometimes, however, a group of resources may hardly be restrictive. This may be true for the cutting tools in a *Flexible Manufacturing Cell (FMC)*. Usually, an FMC consists of a parallel machine group and a large set of unique tools that can only be used by the machines of the FMC. If an FMC is part of the job shop, then the decomposition of this shop results for the FMC in a scheduling problem of minimizing maximum lateness for a parallel-machine group, subject to release dates and tooling restrictions. The interaction of the tools and the machines in the FMC should be handled by an algorithm for scheduling the FMC, not the SB procedure. This is why we call it the decentralized approach. The schedule of *each* resources, *i.e.*, including the tools, needs to be represented by arcs in the graph G. Note that in contrast to the centralized approach, this approach affects both G and the machine scheduling subproblems.

Meester and Zijm [MZ93] present a hierarchical algorithm to schedule an FMC. In this algorithm, two hierarchical levels are distinguished: the top level at which the unique tools are scheduled and the lower level at which the parallel-machine group is scheduled. The unique tools can perform only one operation at a time and are therefore less flexible in scheduling some specific jobs, whereas the machines, due to their common properties, are fully interchangeable and are therefore more flexible in repositioning preset jobs. Meester and Zijm compare the performance of the algorithm with lower bounds obtained by relaxing the multi-resource constraints. As in the job shop scheduling problem, the gap between the lower and upper bounds is quite large. The authors feel that this is due to the weakness of the lower bounds.

Meester [Mee96] tests both the centralized and the decentralized approach on real-life instances. In one case, he tests the centralized approach in the machine shop of Ergon B.V. in Apeldoorn (The Netherlands), where each operation needs also an operator during processing. In another case, Meester tests the decentralized approach in the machine shop of El-o-Matic B.V. in Hengelo (The Netherlands). This machine shop consists of conventional and *Computer Numerically Controlled* (CNC) machines, among which one FMC with a large number of unique tools. Compared with the planning procedure used by the companies, the SB procedure shows a significant improvement of the due date performance in both cases. The problems encountered at El-o-Matic have served as an important source of inspiration for the development of the scheduling system JobPlanner (*cf.* Zijm [Zij96]).

2.4 Set-up times

A machine may have to be set up before it can process the next operation. This happens, for instance, when tools must be switched off-line or when the machine must be cleaned between two operations. During a set-up, the machine cannot process any operation. This means that setting up a machine means in fact a loss of capacity. A natural way of coping with set-up times on a machine is to schedule consecutively operations with the same set-up characteristics. This means that we use this machine efficiently. This may result, however, in a poor delivery performance for products that have operations with other set-up characteristics. Even shop efficiency may decrease, since other machines do not receive jobs in time. We therefore have to find a trade-off between efficiency and delivery performance.

Suppose that a partial schedule on the machine with set-up times is $O_{gh} - O_{ij}$. The set-up time between O_{gh} and O_{ij} is $s_{gh,ij}$. In the graph, we model this set-up time by giving weight $s_{gh,ij}$ to the arc from v_{gh} to v_{ij}, representing part of the schedule of the machine. The length of a longest path from v_{gh} to v_{ij} in the graph G is then at least $s_{gh,ij}$. This ensures that there are at least $s_{gh,ij}$ time units between the completion of O_{gh} and the start of O_{ij}, which leaves room for the needed set-up.

In the standard SB procedure, we need to solve the single-machine problem $1|r_j|L_{\max}$ a number of times. Now, the set-up times between the execution of the operations need to be taken into account, *i.e.*, we have to solve the $1|r_j, s_{ij}|L_{\max}$ problem. For single-machine scheduling problems with family set-up times, *i.e.*, for the $1|r_j, s_i|L_{\max}$ problem, Schutten *et al.* [SvdVZ96] present a branch-and-bound algorithm that solves instances with up to 40 jobs to optimality in reasonable time. A major algorithmic novelty is the use of set-up *jobs* for lower bounding purposes that represent set-up *times*. Schutten *et al.* provide sufficient conditions when set-up jobs may be derived and specify their release and due dates, processing times, and precedence relations with the normal jobs. An upper bound *ub* on the optimal maximum lateness induces for each job J_j a *deadline* $\bar{d}_j = d_j + ub$ for J_j in any optimal schedule. A key observation is that job sets $\mathcal{A}$ and $\mathcal{B}$ may be separated by a set-up job if the deadlines of jobs in $\mathcal{A}$ are sufficiently smaller than the release dates of jobs in $\mathcal{B}$. As an example, consider the job set $\mathcal{F}_i$, consisting of all jobs of family i, and suppose that $\mathcal{A} \subsetneq \mathcal{F}_i$ and $\mathcal{B} = \mathcal{F}_i \setminus \mathcal{A}$. If $\max_{J_j \in \mathcal{A}} \bar{d}_j + s_i \leq \min_{J_j \in \mathcal{B}} r_j$, then we may introduce a set-up job that separates the sets $\mathcal{A}$ and $\mathcal{B}$. Additional set-up jobs can be derived under relaxed conditions.

Accordingly, the SB procedure can deal with set-up times by changing the length of the arcs in G, and by exploiting an algorithm for the machine scheduling subproblems with set-up times, such as the one mentioned above.

Belderok [Bel93] tests the SB procedure with set-up times in the Sheet Metal Factory of DAF trucks in Eindhoven (The Netherlands). In this factory, set-up times occur when changing the moulds of the presses. His computational experiments indicate that a significant leadtime reduction along with a better due date performance is possible, in particular for the Sheet Metal Press department. For example, Belderok tests the procedure on a real-life set of jobs that were processed in five days. The makespan of the schedule generated by the SB procedure is less than three and a half days.

2.5 Convergent and divergent job routings

In the classical job shop problem, each job is a *chain* of operations. In practice, job routings may be convergent. A *convergent* job routing occurs when some components are assembled. Figure 2.3 shows a representation of an instance with a convergent job routing. An example of a *divergent* job routing is the routing of a metal sheet.

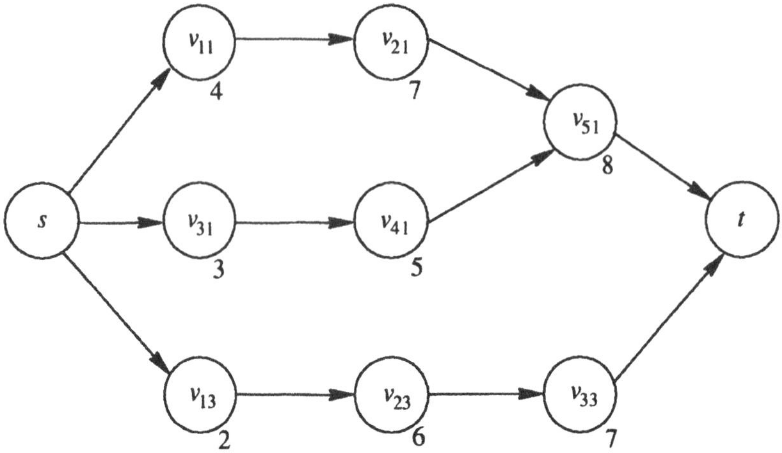

FIGURE 2.3. Instance in which J_1 has a convergent routing.

Before cutting, the sheet needs some operations such as cleaning and surface treatments. After cutting, the different parts of the sheet have their own routings through the shop. We model this by allowing the nodes in G to have more than one ingoing and outgoing job arc. Note that this modeling of convergent and divergent job routings only influences the properties of G, not the machine scheduling subproblems. Schutten [Sch95, Sch96] compares the due date performance of SB procedure to the performance priority rules in an assembly shop. The shop consists of a number of machines producing components and one machine assembling components to end products. On the latter machine, set-up times may occur when switching from assembling an end product to another. The effect of set-up times and the arrival process is studied in this shop. Also, the effect of dynamic scheduling instead of static scheduling is studied. Test results show that the SB procedure significantly outperforms priority rules.

2.6 Further extensions and practical aspects

In this section, we present some further extensions of the basic decomposition approach to deal with practical scheduling problems. We discuss transportation times, unequal transfer and production batches, and open shops.

2.6.1 Transportation times

In practice, it may be impossible to start operation O_{ij} immediately after the completion of operation $O_{i-1,j}$, because the product must first be transported from machine $\mu_{i-1,j}$ to machine μ_{ij}. If the transportation capacity is unlimited, *i.e.*, the transportation of a product always starts immediately after the completion of the operation, then we model this by giving the arc from $v_{i-1,j}$ to v_{ij} a weight that is equal to the transportation time. This creates enough time between the completion of $O_{i-1,j}$ and the start of O_{ij} to transport the product to the next machine. Note that we need not change any algorithm for the machine scheduling subproblems to deal with this type of transportation time.

Reesink [Ree93] tests the SB procedure with transportation times at Stork Plastics Machinery in Hengelo, The Netherlands. He also uses transportation times to model operations that are subcontracted. These operations are assumed to have a fixed leadtime. Belderok [Bel93] uses transportation times to make the resulting schedule more *robust*. A schedule is robust if a small increase in the processing time of an operation does not create the need to reschedule. Frequent rescheduling may lead to nervousness on the shop floor, if the operators have to deal with frequently changing schedules.

2.6.2 UNEQUAL TRANSFER AND PRODUCTION BATCHES

A job may represent an order to produce a batch of b identical products, not just a single product. An operation O_{ij} of this job is then actually a series of b identical operations: $O_{ij} = (O_{i,1,j}, O_{i,2,j}, \ldots, O_{i,b,j})$. If the b identical products need to be processed *contiguously* on each machine, then O_{ij} is called a *production batch*. We assume that a production batch needs to be processed continuously, i.e., without idle time, on the machines. Suppose now that we may transport $O_{i,k,j}$ $(k = 1, \ldots, b-1)$ to the next machine immediately after its completion. If we do this, then it may result in a smaller completion time on the next machine for the production batch. We call $O_{i,k,j}$ a *transfer batch*. For problems with $p_{ij} > p_{i+1,j}$, we shift the batches on machine $\mu_{i+1,j}$ to the right, such that no idle time between the batches on this machine exists; see Figure 2.4 for an example with $b = 4$. Note that the difference in

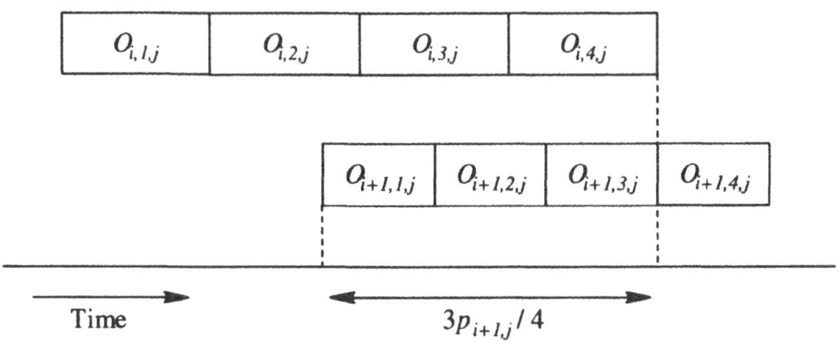

FIGURE 2.4. Transfer batches with $p_{ij} > p_{i+1,j}$ and $b = 4$.

time between the completion of O_{ij} and the start of $O_{i+1,j}$ is at least $-\frac{b-1}{b} \cdot p_{i+1,j}$ time units. For problems with $p_{ij} \leq p_{i+1,j}$, the transfer batches may immediately be processed on the next machine after transporting it; see Figure 2.5. The difference between the start of $O_{i+1,j}$ and the completion of O_{ij} is now at least $-\frac{b-1}{b} \cdot p_{ij}$ time units. The SB procedure can therefore deal with unequal transfer and production batches if we give arc from v_{ij} to $v_{i+1,j}$ weight $-\frac{b-1}{b} \cdot \min\{p_{ij}, p_{i+1,j}\}$.

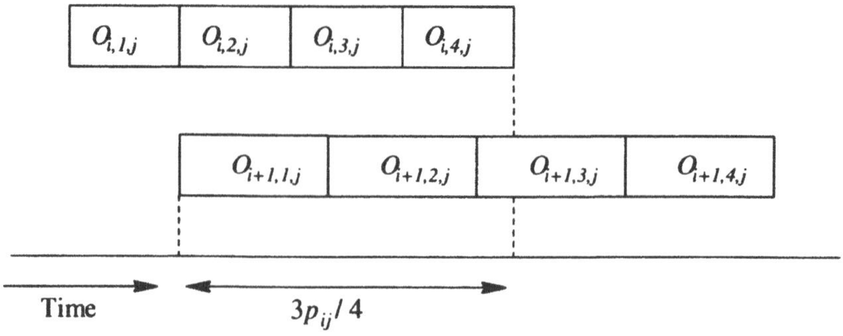

FIGURE 2.5. Transfer batches with $p_{ij} \leq p_{i+1,j}$ and $b = 4$.

2.6.3 Open shops

In the classical job shop, the sequence in which the operations of a job must be processed is given. In open shop problems, it is not: the operations can be performed in any order, although the operations of the same job cannot be processed simultaneously. We model this by introducing for each job a single, artificial machine on which the operations of this job must be processed. The schedule on the artificial machine dictates the sequence in which the operations of the corresponding job are processed. Each operation needs two resources: the artificial machine and the machine on which the actual processing takes place. Also, we need arcs from s to v_{ij} $(j = 1, \ldots, n; i = 1, \ldots, n_j)$ to ensure a path from s to every other node. Analogously, we need an arc from v_{ij} to t to ensure a path from node v_{ij} to t. Figure 2.6 shows the disjunctive graph model of an open shop problem with $n = m = 2$, the data of which are found in Table 2.2. The solid edges in the figure indicate that O_{11}

J_j	μ_{1j}	μ_{2j}	p_{1j}	p_{2j}
J_1	1	2	7	9
J_2	1	2	3	5

TABLE 2.2. Open shop problem.

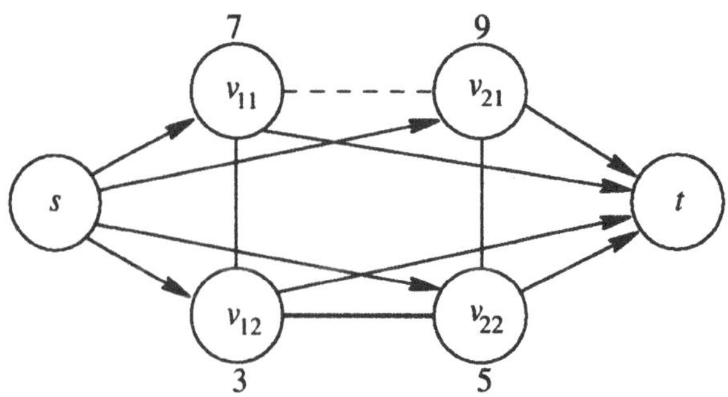

FIGURE 2.6. Representation of the instance of the open shop problem.

and O_{12} as well as O_{21} and O_{22} need to be processed on the same machine, and therefore cannot be processed simultaneously. O_{11} and O_{21} as well as O_{12} and O_{22} also cannot be processed simultaneously, because they are operations of the same job. This is indicated by the non-solid edges.

Once having introduced artificial machines as additional resources, the procedures developed by Meester [Mee96] (see also Section 2.3) can be applied to solve open shop problems.

2.7 JobPlanner

Based on the decomposition framework and the algorithms outlined in the preceding sections, the commercial software package JobPlanner has been developed and is now operating in practice. Currently, a few other scheduling systems with similar practical features have been developed, see, *e.g.*, Ivens and Lambrecht [IL96] and Ovacik and Uzsoy [OU97]. To the best of our knowledge, all these systems are based on the Shifting Bottleneck procedure. Instead of comparing JobPlanner to another (but similar) system, we will report on practical experiences and show that a modern scheduling system like JobPlanner may be beneficial for companies.

Figure 2.7 presents the communication interfaces between JobPlanner and its environment (*cf.* Heerma and Lok [HL94]). A *production planning system*, often

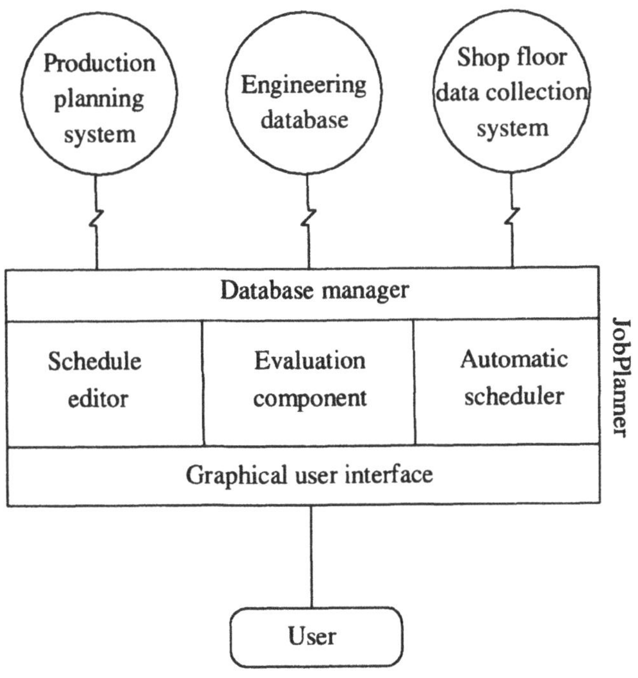

FIGURE 2.7. JobPlanner and its environment.

an MRP II system, informs JobPlanner *which* jobs need to be scheduled. The *engineering database* contains information on *how* a product is produced, *i.e.*, the process planning data. The *shop floor data collection system* monitors the shop status and gathers short term scheduling information.

As displayed in Figure 2.7, JOBPLANNER consists of five main components: the database manager, the automatic scheduler, the graphical user interface, the schedule editor, and the evaluation component. We briefly discuss these components and practical experiences with JOBPLANNER in a manufacturing shop of printed circuit boards.

2.7.1 COMPONENTS OF JOBPLANNER

The database manager communicates with the production planning system, the engineering database, and the shop floor data collection system. It extracts data from these databases and stores it in a *local shop floor database*. In addition to the logistic and engineering information, it contains data about the shop status. For example, when an operation is completed or a machine breaks down, this information is stored in the local shop floor database. This makes this database *dynamic*, *i.e.*, it changes over time.

The automatic scheduler is based on the SB decomposition, enhanced with the algorithms discussed in the preceding sections. Via the database manager, it receives information on job routings and processing times from the local shop floor database. The scheduler proposes a schedule, which is stored in the database.

The graphical user interface is an important means for the interaction between man and machine. The graphical user interface enables the planner to perform actions like starting the automatic scheduler, downloading information from a database, printing of information, and so on. An element in the graphical user interface for scheduling systems is the electronic planning board. The electronic planning board uses Gantt charts (Gantt [Gan19]) to represent schedules and provides the means to modify schedules by manipulating the Gantt chart; *cf.* Wennink [Wen95].

Besides a graphical representation, a scheduler needs data to evaluate the quality of a schedule. Useful is, *e.g.*, data on the delivery performance, machine utilization, makespan, and throughput times. Data can be represented in the form of tables, graphs, throughput diagrams (see Wiendahl [Wie87]), and so on.

In a scheduling system, the planner should have the ability to create, modify, store, and retrieve schedules. The scheduler oversees, for example, practical problems that are not covered by the underlying models. The *interaction* between the planner and the scheduling system is crucial to end up with a feasible schedule. In this context, interaction means the integration of human perception and mechanical algorithms; *cf.* Savelsbergh [Sav88].

2.7.2 PRACTICAL EXPERIENCES WITH JOBPLANNER

JOBPLANNER and its algorithms have been tested in various companies in discrete manufacturing shops, see Belderok [Bel93], Reesink [Ree93], and Wilbers [Wil94]. In this section, partly based on Heerma [Hee], we report on experiences with JOBPLANNER at Cityprint B.V., located in Almelo, The Netherlands.

Cityprint is a Dutch producer of printed circuit boards (PCBs) with currently 140 employees, of which about 100 are production personnel or production engineers.

The machine shop consists of 40 machines. The annual turnover is more than 25 million guilders, of which more than 50% is due to export to countries in Europe. Production, and even engineering, is to customer order. The leadtimes are at most two to four weeks. Cityprint uses two base materials for producing PCBs: teflon and epoxy. Teflon is used, for example, for PCBs in satellite dishes and epoxy for PCBs in computers and television sets. The PCBs have one up to eight track layers.

The production system of Cityprint can be divided in six production groups: production engineering, cutting, through hole plating, imaging, pressing, and testing. For each customer order, *production engineering* launches an internal order. For each order, it determines the necessary production steps, selects the machines on which they have to be performed, and the machining order. The *cutting* group cuts teflon and epoxy plates covered with copper foil to the required sizes and drills holes in them. The *through hole plating* group takes care that the faces of the drilled holes get plated with copper. Whenever there is a switch from plating teflon PCBs to plating epoxy PCBs and vice versa, a set-up time is needed. The *imaging* group adds the copper tracks to the PCBs. It uses a photo-sensitive material and a film of the track. At places that have not been exposed to light, the copper is removed through etching. The *pressing* group presses a number of plates to one multi-layered PCB. Finally, the *testing* group tests the produced PCBs for example on short-circuits. A typical job consists of 20 operations.

The introduction of JOBPLANNER at Cityprint has been satisfactory. First of all, the graphical representation of schedules has increased the insight of the planner into the actual workload of the production system, which has made it relatively easy to trace bottlenecks in the production system and to respond appropriately. The sales department can also use the insight in the actual workload of the production system by quoting feasible and competitive delivery dates. The introduction of JOBPLANNER has reduced the time for production planning. It used to be a full-time job; now, it takes only one or two hours a day. Hence, there is more time to work out structural improvements of the production process. In the first year that JOBPLANNER was used, 95% of the jobs were delivered in time, the output increased by 15%, and the throughput times decreased by 25%. When this paper was going to press, two other factories have adopted JOBPLANNER. Up to now, results for those factories cannot be reported.

2.8 Conclusions

In this paper, we motivated the application of advanced planning and scheduling systems in discrete parts manufacturing. An example of such a system is presented, based on the Shifting Bottleneck decomposition and enhanced with a number of algorithms to deal with complex, but realistic, manufacturing systems. Experiences with the system in practice yield highly satisfactory results. In summary, the application of intelligent scheduling systems for realistic systems is needed, possible, and worthwhile.

2.9 REFERENCES

[ABZ88] J. Adams, E. Balas, and D. Zawack. The Shifting Bottleneck procedure for job shop scheduling. *Management Science*, 34:391–401, 1988.

[Bec87] W. Bechte. Theory and practice of load-oriented manufacturing control. *International Journal of Production Control*, 26:375–396, 1987.

[Bel93] G.A. Belderok. Ontwerp van een produktiebesturings- en informatiesysteem voor de produktie van zware persdelen in de plaat komponenten fabriek van DAF Trucks Eindhoven (in Dutch). Master's thesis, University of Twente, Faculty of Mechanical Engineering, Production and Operations Management Group, Enschede, The Netherlands, 1993.

[BJS94] P. Brucker, B. Jurisch, and B. Sievers. A branch and bound algorithm for the job-shop scheduling problem. *Discrete Applied Mathematics*, 49:107–127, 1994.

[Bla91] J.D. Blackburn. *Time-Based Competition, The Next Battle Ground in American Manufacturing*. Richard D. Irwin, Homewood, Ill., 1991.

[BW81] J.W.M. Bertrand and J.C. Wortmann. *Production Control and Information Systems for Component Manufacturing Shops*. Elsevier, Amsterdam, 1981.

[Car82] J. Carlier. The one-machine sequencing problem. *European Journal of Operational Research*, 11:42–47, 1982.

[CP89] J. Carlier and E. Pinson. An algorithm for solving the job-shop problem. *Management Science*, 35:164–176, 1989.

[CTV95] F. Della Croce, R. Tadei, and G. Volta. A genetic algorithm for the job-shop scheduling problem. *Computers and Operations Research*, 41:231–252, 1995.

[Dem82] W.E. Deming. *Quality, Productivity, and Competitive Position*. MIT Center for Advanced Engineering Study, Cambridge, Mass., 1982.

[DH96] E.L. Demeulemeester and W.S. Herroelen. Modeling setup times, process batches and transfer batches using activity network logic. *European Journal of Operational Research*, 89:355–365, 1996.

[FT63] H. Fisher and G.L. Thompson. Probabilistic learning combinations of local job-shop scheduling rules. In J.F. Muth and G.L. Thompson, editors, *Industrial Scheduling*, pages 225–241. Prentice-Hall, Englewood Cliffs, NJ, 1963.

[Gan19] H.L. Gantt. *Organizing for Work*. Harcourt, Brace and Howe, New York, 1919.

[Glo89] F. Glover. Tabu search - Part I. *ORSA Journal on Computing*, 1:190–206, 1989.

[Glo90] F. Glover. Tabu search - Part II. *ORSA Journal on Computing*, 2:4–32, 1990.

[Gol88] E.M. Goldratt. Computerized shop floor scheduling. *International Journal of Production Research*, 26:443–455, 1988.

[HC93] M. Hammer and J. Champy. *Re-engineering the Corporation: A Manifesto for Business Revolution*. Harper Collins, 1993.

[Hee] W. Heerma. Jobplanner, Gereedschap voor betere logistieke prestaties (in Dutch). To appear in: *Via's, Magazine for the Electrotechnical Industry.*

[HL94] W. Heerma and N.R. Lok. Scheduling-systeem met IQ ontlast planner (in Dutch). *Logistiek Signaal*, 4:27–29, 1994.

[HM75] A.C. Hax and H.C. Meal. Hierarchical integration of production planning and scheduling. In M.A. Geisler, editor, *Logistics (Studies in the Management Sciences, Vol. 1)*. Elsevier, North-Holland, 1975.

[HW89] N.L. Hyer and U. Wemmerlov. Group technology in the U.S. manufacturing industry: A survey of current practices. *International Journal of Production Research*, 27:1287–1304, 1989.

[IL96] P. Ivens and M. Lambrecht. Extending the shifting bottleneck procedure to real-life applications. *European Journal of Operational Research*, 90:252–268, 1996.

[LAL92] P.J.M. van Laarhoven, E.H.L. Aarts, and J.K. Lenstra. Job shop scheduling by simulated annealing. *Operations Research*, 40:113–125, 1992.

[LLK82] E.L. Lawler, J.K. Lenstra, and A.H.G. Rinnooy Kan. Recent developments in deterministic sequencing and scheduling: A survey. In M.A.H. Dempster, J.K. Lenstra, and A.H.G. Rinnooy Kan, editors, *Deterministic and Stochastic Scheduling*, pages 35–73. NATO Advanced Study and Research Institute, D. Reidel Publishing Company, Dordrecht, The Netherlands, 1982.

[Mee96] G.J. Meester. *Multi-Resource Shop Floor Scheduling*. PhD thesis, University of Twente, Enschede, The Netherlands, 1996.

[MP93] T.E. Morton and D.W. Pentico. *Heuristic Scheduling Systems: With Applications to Production Systems and Project Management*. John Wiley & Sons, New York, 1993.

[MZ93] G.J. Meester and W.H.M. Zijm. Multi-resource scheduling for an FMC in discrete parts manufacturing. In M.M. Ahmad and W.G. Sullivan, editors, *Flexible Automation and Integrated Manufacturing*, pages 360–370. CRC Press Inc., Atlanta, 1993.

[Orl75] J. Orlicky. *Material Requirements Planning*. McGraw-Hill, New York, 1975.

[OU97] I.M. Ovacik and R. Uzsoy. *Decomposition Methods for Complex Factory Scheduling Problems.* Kluwer Academic Publishers, 1997.

[Pin95] M. Pinedo. *Scheduling: Theory, Algorithms, and Systems.* Prentice Hall, Englewood Cliffs, 1995.

[Ree93] F.F.J. Reesink. Werkplaatsbesturing met behulp van de Shifting Bottleneck Methode (in Dutch). Master's thesis, University of Twente, Faculty of Mechanical Engineering, Production and Operations Management Group, Enschede, The Netherlands, 1993.

[RS64] B. Roy and B. Sussman. Les problèmes d'ordonnancement avec contraintes disjonctives. Note DS No. 9 bis, SEMA, Paris, 1964.

[Sav88] M.W.P. Savelsbergh. *Computer Aided Routing.* PhD thesis, Erasmus University, Rotterdam, 1988.

[Sch] J.M.J. Schutten. Practical job shop scheduling. To appear in: *Annals of Operations Research.*

[Sch82] R.J. Schonberger. *Japanese Manufacturing Techniques.* The Free Press (MacMillan), New York, 1982.

[Sch95] J.M.J. Schutten. Assembly shop scheduling. Technical Report LPOM-95-15, University of Twente, Faculty of Mechanical Engineering, Production and Operations Management Group, Enschede, The Netherlands, 1995.

[Sch96] J.M.J. Schutten. *Shop Floor Scheduling with Setup Times: Efficiency versus Leadtime Performance.* PhD thesis, University of Twente, Enschede, The Netherlands, 1996.

[SvdVZ96] J.M.J. Schutten, S.L. van de Velde, and W.H.M. Zijm. Single-machine scheduling with release dates, due dates and family setup times. *Management Science*, 42:1165–1174, 1996.

[Tie92] J.J. Tiemersma. *Shop Floor Control in Small Batch Part Manufacturing.* PhD thesis, University of Twente, Enschede, The Netherlands, 1992.

[VAL96] R.J.M. Vaessens, E.H.L. Aarts, and J.K. Lenstra. Job shop scheduling by local search. *INFORMS Journal on Computing*, 8:302–317, 1996.

[Wen95] M. Wennink. *Algorithmic Support for Automated Planning Boards.* PhD thesis, Eindhoven University of Technology, Eindhoven, The Netherlands, 1995.

[Wie87] H. Wiendahl. *Belastungsorientierte Fertigungssteuerung.* Carl Hanser, Munich, 1987.

[Wig84] O.W. Wight. *Manufacturing Resource Planning: MRP II.* Oliver Wight Ltd., Essex Junction, 1984.

[Wil94] A.A.B. Wilbers. Capaciteitsplanning bij Urenco Nederland B.V. (in Dutch). Master's thesis, University of Twente, Faculty of Mechanical Engineering, Production and Operations Management Group, Enschede, The Netherlands, 1994.

[WJR90] J.P. Womack, D.T. Jones, and D. Roos. *The Machine that Changed the World.* Maxwell MacMillan International, New York, 1990.

[Zij96] W.H.M. Zijm. Shopfloor planning and control for small-batch parts manufacturing. In L. Fortuin, P. Van Beek, and L.N. Van Wassenhove, editors, *OR at Work*, chapter 7, pages 111–132. Taylor & Francis, 1996.

Appendix A: Derivation of set-up jobs

Consider any instance I of $1|r_j, s_i|L_{\max}$ and let I' be the instance of $1|r_j, \textit{setup-prec}|$ $L_{\max}$ obtained from I by ignoring the family set-up times. Let $L^*_{\max}(I)$ ($L^*_{\max}(I')$) denote the optimal solution value for instance I (I'). We have that $L^*_{\max}(I') \leq L^*_{\max}(I)$, since we ignore the set-up times in I'. Suppose now that we have established, one way or the other, that in every optimal schedule for I all jobs in $\mathcal{A} \subset \mathcal{F}_i$ precede all jobs in $\mathcal{B} \subset \mathcal{F}_i$ ($\mathcal{B} \neq \emptyset$) and no job from $\mathcal{A}$ and no job from $\mathcal{B}$ are scheduled in the same batch, where a *batch* is a set of jobs that are scheduled between two subsequent set-ups. This then means that there must be at least one *separating* set-up associated with family $\mathcal{F}_i$ between the last job belonging to $\mathcal{A}$ and the first job belonging to $\mathcal{B}$. The next theorem validates our key idea that this setup can be viewed as a *separating setup job* with a specific processing time, release date, due date, and precedence relations. Let $\succ$ and $\prec$ mean 'has to follow' and 'has to precede', respectively.

Theorem We still have that $L^*_{\max}(I') \leq L^*_{\max}(I)$, if we add a setup job J_s to I' with

$$\begin{aligned} p_s &= s_i, \\ J_s &\succ J_j, \text{ for all } J_j \in \mathcal{A}, \\ J_s &\prec J_j, \text{ for all } J_j \in \mathcal{B}, \\ r_s &= \min_{J_j \in \mathcal{F}_i \setminus \mathcal{A}} r_j - s_i, \\ d_s &= \min_{J_j \in \mathcal{B}} (d_j - p_j). \end{aligned}$$

Proof It only remains to be shown that the specification of r_s and d_s is correct. Consider any optimal schedule σ for I and any setup for family $\mathcal{F}_i$ that succeeds all jobs from $\mathcal{A}$ and precedes all jobs from $\mathcal{B}$ in this schedule. We associate the setup job J_s with this setup. We may assume that this setup occurs immediately before the execution of the job it is needed for. Since this may be any job in $\mathcal{F}_i \setminus \mathcal{A}$, the release date of J_s follows. Let σ' be the feasible schedule for I' obtained from σ in the following way: let the sequence of the real jobs in σ' concur with the sequence in σ, and replace the setups with their associated setup jobs, if they have one. Note that $C_j(\sigma') \leq C_j(\sigma)$ for all real jobs (*i.e.*, not the set-up jobs), and therefore

$L_j(\sigma') \leq L_j(\sigma) \leq L^*_{\max}(I)$. If we assign d_s as proposed, we have that $d_s = d_j - p_j$ and $J_s \prec J_j$ for some $J_j \in \mathcal{B}$, and hence that

$$\begin{aligned} L_s(\sigma') &= C_s(\sigma') - d_s \leq C_j(\sigma') - p_j - (d_j - p_j) \\ &\leq C_j(\sigma) - d_j = L_j(\sigma) \leq L^*_{\max}(I). \end{aligned}$$

Thus, we proved that $L_j(\sigma') \leq L^*_{\max}(I)$ for every job in I', and therefore $L^*_{\max}(I') \leq L_{\max}(\sigma') \leq L^*_{\max}(I)$. □

The crux is that the addition of this separating setup job may improve the value $L^*_{\max}(I')$, and thus the lower bound on $L^*_{\max}(I)$.

3

Integrating Layout Design and Material Flow Management in Assembly Systems

M. Lucertini[1]
D. Pacciarelli[2]
A. Pacifici[2]

ABSTRACT The material flow management of flexible assembly systems requires to assign a set of operations to a set of machines, and to connect machines by a transportation network, such that a number of constraints is satisfied and some efficiency index is optimized. Aim of the paper is to give a general framework to formulate and model, in a formal way, different subproblems arising from embedding an assembly process on different configurations of a flexible production system. We analyze several wide used architectures and material flow policies, and point out some useful properties to design decision rules and evaluate the corresponding optimality bounds. The paper also investigates the computational complexity of various subcases of the problem.

3.1 Introduction

Flexible Assembly Systems (FAS) are a class of automated systems which can be used to improve productivity in discrete manufacturing.

In its basic form an *assembly process* consists on assemblying raw parts and performing all the related operations in order to produce a finished product. (For some contributions on manufacturing models and assembly systems see, for example, [3].) We call *operation* the smallest work element whose execution cannot be interrupted and must be performed by a single machine. Each operation requires the availability of some component subassemblies and produces one or more resulting subassemblies. Therefore it cannot be performed before all the operations producing the component subassemblies are completed. This fact induces a set of *precedence constraints* among operations. Finally, for each operation a *processing time* is given. Throughout this chapter we will consider the processing times as fixed and not depending on the performing machine.

When an assembly process is embeddded in a plant layout, parts move from one machine to another via a set of transportation connections. The material handling

---

[1]Dipartimento di Informatica, Sistemi e Produzione, Università "Tor Vergata", Roma, Italy
[2]Dipartimento di Informatica e Automatica, Università di Roma Tre, Roma, Italy

system can be modeled as a set of positions where machines can be placed in, and a set of connections among positions.

We will consider completely reliable systems (machines and transportation network never fails and parts are always avalilable) dedicated to the production of a single finished product.

Due to the complexity of coordinating these systems, with all its different types of constraints (placement, transportation, size, weight, information flow, computational, etc.), even very simple architectures may produce complex material flow patterns. Usual efficiency indicators (cycle time, completion time, work in process, workload, number of part transfers etc.) become difficult to be evaluated and even to be formulated in logical terms or as the optimal solution of a decision problem expressed in analytical form.

The problem addressed in this chapter is the following: given an assembly process, a set of machines and a trasportation network, find the assignment of operations to a set of available machines, the assignment of chosen machines to physical positions, the routing of the units among machines, and the sequencing of operations on each machine, such that a suitable objective function is minimized. Several relevant problems in production optimization can be viewed as special cases of this problem. Among the others, machine loading (see [15, 25]), machine layout (see [17, 19]), assembly line balancing, and several routing and scheduling problems (see [9]).

In particular assembly line balancing (ALB) is a traditional problem in industrial engineering and it has been the subject of a great number of papers, since the first formulation by Hengelson et al. in 1954 [18]. ALB falls into the class of NP-hard problems [13], therefore most of the research efforts has been made to develope efficient heuristics solving large scale ALB problems. See for example [4, 12]. As shown in the Chase's survey [10] and more recently in [20], *"in spite of hundreds of works on assembly lines, only a little number of companies utilizes published techniques to balance their lines"*. Nonetheless, there are celebrated examples of companies which were able to enhance productivity by means of innovative methods, such as [23].

Actually, most of the models usually adopted for material flow management in assembly systems suffer from substantial loss of information. In fact, the trasportation network may differ from a line, and little work has been done in this case on modeling the full range of factors affecting productivity. Moreover, the most common performance indices adopted in literature concern makespan and cycle time, whereas other factors have also proved to be effective with reference to production optimization in some manufacturing systems: the minimization of the number of part transfers, i.e. the number of subassemblies moving from a machine to another, corresponds to a surrogate for taking into account both traffic problems and set-up costs [21, 22]. This performance index did not receive much attention in literature, so far. In fact, this objective is not significant if transportation and set-up times are small with respect to the processing times, i.e. when the transportation network does not introduce delays due to the traffic congestion problems. On the other hand, minimizing the part movements can be better than balancing the machine workloads whenever the transportation device is a system bottleneck [24].

In order to reduce transportation network congestion, it can also be profitable to avoid the existence of particular structures in the material flow pattern that may affect system productivity, such as cycling of parts in the system, i.e., parts entering the same machine at different times of the production process.

In this chapter we propose a general framework to formulate and model, in a precise way, different subproblems arising from embedding an assembly process on different layout configurations of a flexible production system. In section 3.2 we introduce the notation and a numerical example. Then we give a formal statement of the problem in an optimization format. In section 3.3, we will present some general properties concerning the feasibility of a solution, given certain partial decisions. In section 3.4 we deal with the problem of optimally assigning operations to physical machines located in specified positions.

3.2 Statement of the problem

Hereafter we give a formal statement of the problem in an optimization format.

3.2.1 PROBLEM DATA

Let N be the set of n operations of the assembly process, and A be the corresponding set of precedence constraints. Then the assembly process can be represented as a weighted acyclic graph $G = (N, A)$ (in the following *Assembly Graph*), usually a tree, where the node weights correspond to the operations' processing times. If an arc $(i, j) \in A$ then a machine k is allowed to perform operation j only when the subassembly resulting from operation i is available on machine k. In this paper it is assumed that, if no precedence relationship exists between i and j, these operations are performed on different subassemblies and can be executed in parallel.

The operations must be performed by a set of machines chosen in a set M of available machines. A machine $k \in M$ can execute only one operation at a time, preemption is not allowed. We assume that the machine tooling is decided *a priori* and no reconfiguration is allowed during the running of the production process, this case is referred to in literature as *static tool allocation*, see for example [8]. Therefore each machine can perform a fixed subset of all the operations.

Plant layout, i.e., physical connections among positions, can be represented by a a directed graph $E = (P, C)$, denoted as *transportation network*, where P is a set of available positions in the plant and C is the set of physical connections between positions. An arc $(i, j) \in C$ if and only if there is a physical connection from position i to position j.

Observe that we do not make any assumption on the numbers of machines and positions available. In fact, in different applications of our model these numbers could be very different from each other. For example in Machine Layout models the number of positions available for machines can be greater than the number of machines (see [5]), whereas in Tooling problems the number of possible configurations for machines usually by far exceeds the number of physical positions in the plant (see [2]). In the remainder of the chapter we denote by m the actual number of machines and positions used (of course $m \leq \min\{|M|, |P|\}$, where $|X|$ indicates the number of elements of set X.) Observe that m can be stricly smaller than $\min\{|M|, |P|\}$. As the mention of the fact, for example, in tooling problems m is the actual number of machines available in the plant, whereas M is the set of all the possible machine configurations and m can be smaller than the number of available positions in the plant.

3.2.2 DECISION VARIABLES

The assignment of operations to machines (*loading*) is equivalent to: (i) select m machines in the set M and (ii) assign the operations to the selected machines. These two decisions result in a partition $\lambda = \{S_1, S_2, \ldots S_m\}$ of N, where subset S_k in the partition contains all the operations to be performed by the machine associated to S_k, $k = 1, \ldots, m$. Clearly, λ is feasible if and only if all the operations in S_k can be performed by the machine associated to S_k.

Physical placement of machines (*positioning*) consists on assigning the selected machines to m physical positions of the plant. An assignment $\pi : M \to P$ is feasible if each machine is compatible with its position (since there may be, in general, forbidden pairs machine-position.) We will denote by $\pi_i \in P$, $i \in N$, the position assigned to the machine that performs operation i.

System performances also depends on two additional decisions, namely the *routing* ρ of subassemblies on the transportation network and the *sequencing* σ of the operations on each machine. More formally, ρ associates to each arc $(i,j) \in A$, such that $\pi_i \neq \pi_j$, the sequence of machines visited by the subassembly resulting from operation i, moving from π_i to π_j, if such a sequence exists. Otherwise, if no such a path exists, ρ associates the empty path to $(i,j) \in A$. The sequencing σ produces, for each machine, a linear ordering of the operations of each subset S_k of λ, for $k = 1, \ldots m$.

It is useful to notice that in most of the applications the assembly graph is an in-tree (i.e., a tree where each node but one has exactly one out-going arc), since any node with more than one out-going arc represents a disassembly operation. In this case, additional constraints on the decisions can be conveniently introduced by imposing $\lambda = \{S_1, S_2, \ldots S_m\}$ to be a *connected partition*, i.e., a partition such that each subgraph induced by S_k is a tree. This implies that each machine produces only one subassembly, possibly the finished product. Observe that, in general, there may be no feasible connected partitions, due to the limited compatibility among operations and machines. On the other hand, if a feasible connected partition is chosen, then the WIP and the number of part transfers is strongly reduced and machine set-up is simplified.

3.2.3 A NUMERICAL EXAMPLE

Let us consider the assembly tree $G = (N, A)$ in Figure 3.1. The 11 operations $a, b, \ldots k$ (with the relative processing times) must be performed on three identical machines M_1, M_2, M_3 which can perform all the operations. The transportation network $E = (P, C)$ is depicted in Figure 3.2.

A solution minimizing makespan is the one shown in Figure 3.3 (solution 1). Here we have $\lambda = \{\{c, d, h, k\}, \{a, e, j\}, \{b, f, g, i\}\}$ (each subset of the partition λ is represented by a color). π assigns machine M_i to position P_i, for $i = 1, 2, 3$. Function ρ trivially associates to each arc $(i,j) \in A$ the arc $(\pi_i, \pi_j) \in C$ (that in this case always exists). Function σ is depicted in the Gantt chart in Figure 3.3. Part transfers

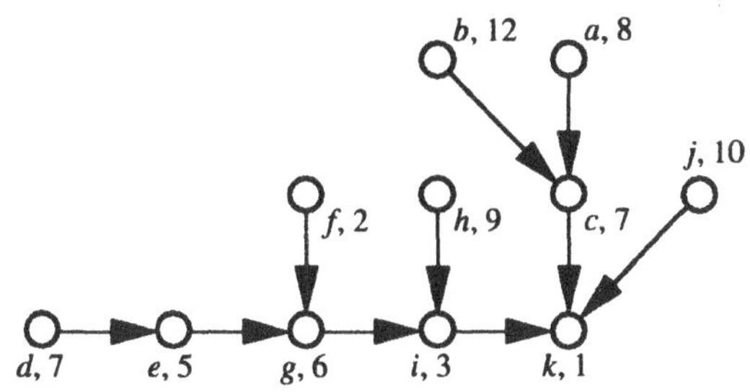

FIGURE 3.1. The assembly tree

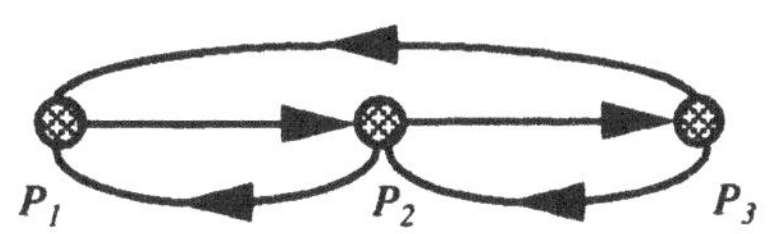

FIGURE 3.2. The transportation network

are highlighted by gray bold lines. In this case the makespan and the cycle time are equal to 24, while the number of part transfer is 7.

Observe that some parts re-enter a previously visited machine. For instance, consider the sequences of operations (h, i, k) and (d, e, g, i, k) which correspond to the sequences of machines (M_1, M_3, M_1) and (M_1, M_2, M_3, M_1), respectively. Clearly the subassemblies associated to operations h and d respectively visit machine M_1 twice. In the following, we will refer to this fact with the term *part cycling*.

It is easy to verify that all the minimum makespan solutions require part cycling. The case of a solution not requiring part cycling is depicted in Figure 3.4. In this case the makespan is 25. The corresponding cycle time remains minimum, equal to 24. The number of part transfers is 4, less than in the previous case. Notice that, by increasing the makespan by 1 (from 24 to 25) and without modifying the throughput, we can reduce the number of part transfers to 4 (solution 2).

In Figure 3.5 an assignment minimizing the number of part transfers is depicted (solution 3).

In this case the cycle time increases to 27, makespan is 28, whereas the number of part transfer is lowered to 2. The numerical results for the above example are summarized in table 3.1:

As shown in table 3.1, the solutions minimizing only one performance index are

Instance	Makespan	Cycle time	Part transfers	Part cycling
Solution 1	24	24	7	yes
Solution 2	25	24	4	no
Solution 3	28	27	2	no

TABLE 3.1. Performance of different solutions

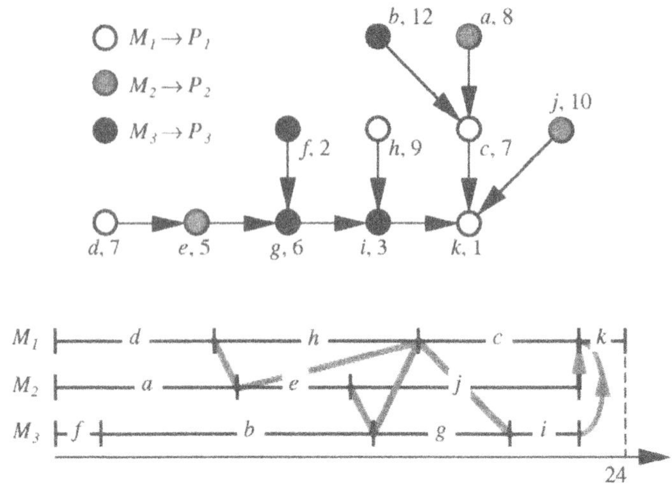

FIGURE 3.3. Solution 1

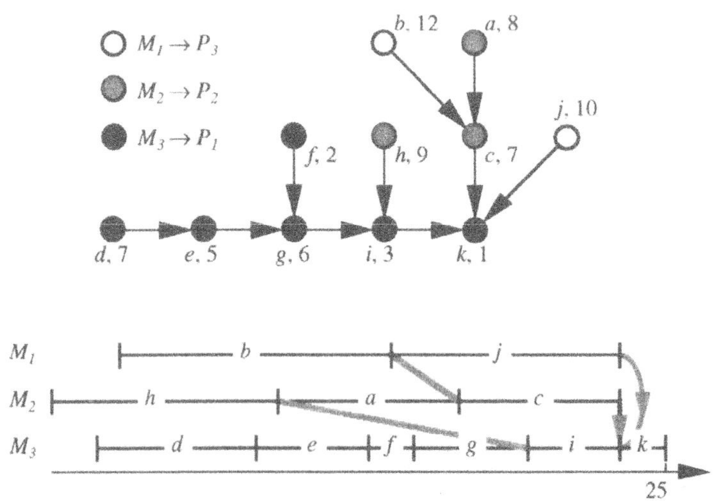

FIGURE 3.4. Solution 2

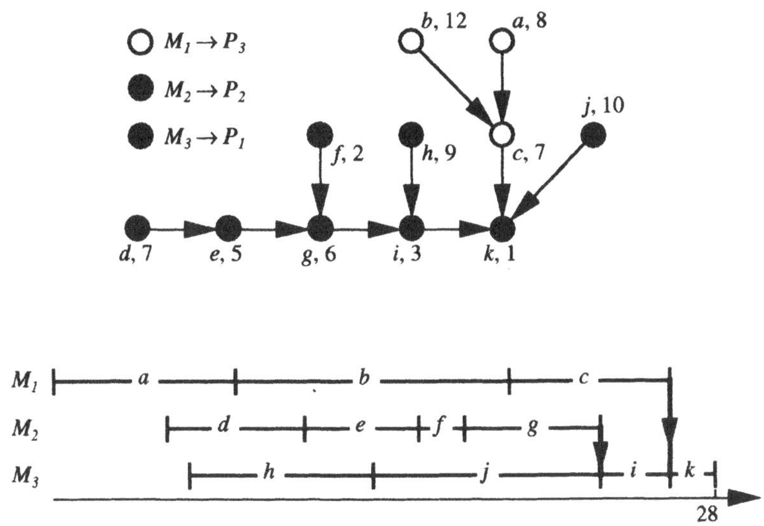

FIGURE 3.5. Solution 3

not efficient with respect to the other, in particular the minimum makespan solution (solution 1) leads to the necessity of part cycling, which may cause congestion in the transportation network. On the other hand, the solution minimizing part transfer (solution 3) is not efficient with respect to the production rate.

3.2.4 PROBLEM STATEMENT

Hereafter, the statement of the problem is given:

Problem 1 (General Problem)

Given : *an assembly graph $G = (N, A)$, a transportation network $E = (P, C)$, a set M of machines, all compatibilities among operations, machines and positions, and an integer m;*

Find : *m machines in the set M, an assignment λ of operations to the m machines, an assignment π of machines to positions, a routing ρ of parts on the transportation network, and a sequencing σ of operations assigned to each machine;*

Such that : *(i) the solution is feasible, i.e. each operation is assigned to a compatible machine, each machine is assigned to a compatible position, routing and sequencing do not violate the precedence relationships expressed by the graphs $E = (P, C)$ and $G = (N, A)$ respectively. Finally, the resulting overall solution is feasible, i.e. the single decisions are compatible each other. (ii) System productivity is maximized.*

Problem 1 generalizes a wide set of decision problems with different levels of aggregation and with different time-span. In practice, the conditions allowing to solve the whole problem within an acceptable computation time are seldom verified. Usually, in a hierarchical approach, problem 1 is solved by considering only a subset of decision variables at a time, while the others are either fixed or considered at a lower hierarchical level in a next stage of the solution procedure.

Depending on the aggregation level (i.e. the subset of variables taken into account) it is possible to address several models, well known in production management, as special cases of problem 1.

1. *Machine loading.* This problem consists of assigning operations to m compatible machines. In this case, λ is the only decision to be taken, while π is given and (ρ, σ) are considered *a posteriori*. The most common objectives addressed in literature are workload balancing and tooling cost minimization [15, 25].

2. *Machine Layout.* Here, the problem consists of determining machine arrangement on the facility floor. In this case, π is the only decision to be taken, while λ is given and (ρ, σ) are considered *a posteriori*. The transportation system is given and the problem is to assign clusters of operations, corresponding to existing machines, to given positions of the plant [17, 19].

3. *Routing and Scheduling.* Here, (ρ, σ) are the only decisions to be taken, while (λ, π) are given. A wide class of well known scheduling problems can be seen in this context. By suitably varying λ and π we can formulate job-shop, flow-shop or more general problems (See, for example, [9]).

3.3 Feasibility properties

In this section we address the problem of finding a feasible solution of 1.

Next we will show that it is possible to represent the effects of the decisions λ (loading), π (positioning), ρ (routing), and σ (scheduling), by suitably modifing the assembly graph $G = (N, A)$, i.e. by adding precedence constraints (arcs) and dummy operations (nodes). In fact, each time a decision is taken, a new set of constraints is added to the problem data. On this resulting graph several performance indices such as completion time, cycle time, and part transfer number, can be easily evaluated.

3.3.1 Feasibility graph

Let us suppose that λ and π are given. It is straightforward to observe that a routing is feasible if and only if the following condition holds:

Condition 1 *For all $(i, j) \in A$ either $\pi_i = \pi_j$ or there exists a path on E from π_i to π_j.*

We may associate to a *routing decision* ρ a set of arcs and (dummy) nodes to be added to G. Let us call *intermediate machines* all the machines of the sequence (if any) which are different from π_i and π_j. We distinguish three cases:

1. Given an arc $(i, j) \in A$, no path exists from π_i to π_j, in this case there is no feasible part transfer; we can represent this fact by adding arc (j, i) to the graph $G = (N, A)$. Let us denote by L the set of arcs (j, i) obtained considering

each arc $(i,j) \in A$. As a consequence of the definition of L, condition (1) is equivalent to the condition $L = \emptyset$.

2. If $\pi_i = \pi_j$, then no part transfer is needed. This corresponds to add no arcs to G.

3. If $\pi_i \neq \pi_j$ and there is a path from π_i to π_j, then a shortest path between the two positions is chosen: in particular if there is a directed connection then the path is formed by the two end-point positions only. If the path contains more than two machines, we must introduce a new set of $k(i,j)$ dummy operations with zero processing time, assigned to each intermediate machine. If we denote with $x^1_{i,j}, x^2_{i,j}, \ldots x^{k(i,j)}_{i,j}$ the dummy operations associated to the arc (i,j), it is possible to build a new operation graph from $G = (N, A)$, by denoting with N' and A' the new sets of nodes and arcs respectively:

$$N' = \{N \cup x^h_{i,j}, 1 \leq h \leq k(i,j), (i,j) \in A\} \tag{3.1}$$

$$A' = \bigcup_{(i,j) \in A} \begin{cases} (i,j) & k(i,j) = 0 \\ (i, x^1_{ij}), (x^1_{ij}, x^2_{ij}), \ldots (x^{k(i,j)}_{ij}, j) & k(i,j) \neq 0 \end{cases} \tag{3.2}$$

A new partition $\lambda' = \{S'_1, S'_2, \ldots S'_m\}$ is now determined, due to the dummy operations introduced by ρ. It is clear that S'_i is the union of S_i and the set of dummy operations assigned by ρ to machine i.

The sequencing decision σ introduces another set O of precedence constraints between operations assigned to the same machine. These constraints simply correspond to adding new arcs to the graph in such a way that m different linear orderings (one for each machine) are produced, i.e. for each pair (i,j) of operations assigned to the same machine, including the dummy operations, either i preceeds j or viceversa. Notice that not all the sequences lead to a feasible solution.

Eventually, let us denote by *feasibility graph* $G' = (N', D)$ the graph where N' is the new set of nodes, introduced in (3.1), and $D = A' \cup L \cup O$. Let us consider any arc $(i,j) \in D$. We associate to this arc a weight equal to the transfer time from π_i to π_j. Of course, in the case of $\pi_i = \pi_j$ the arc weight will be equal to zero. It is straightforward to verify the following:

Remark 1 *Any given solution (λ, π, ρ, σ) is feasible for problem 1 if and only if $G' = (N', D)$ is acyclic.*

Moreover, several performance indices can be easily evaluated on the feasibility graph G'. For example

- Part transfer number is given by the cardinality of the cut induced by the partition λ' in G'.

- Completion time is given by the longest path (given as the sum of node and arc weights) on G'.

- Cycle time equals the longest path between two nodes assigned to the same machine.

Note that if G' contains cycles the longest path has infinite lenght. A necessary and sufficient condition for G' to be acyclic—and thus for a solution to be feasible—is given in the following theorem.

Theorem 1 *Given any assembly graph $G = (N, A)$, for any feasible loading λ, and feasible positioning π, there exist a routing ρ, and a scheduling σ such that $G' = (N', D)$ is acyclic if and only if the condition (1) is verified.*

In order to find a feasible solution, we are allowed to ignore the decisions ρ and σ. In fact, proof of theorem 1 suggests an algorithm to easily find such decisions, given λ and π satisfying condition (1). This allows us to focus our attention on the decisions π and λ, in particular, we are introducing the following two problems:

Problem 2 [FP_π.] *Given a feasible positioning π, find a feasible loading λ such that condition (1) holds.*

We denote this problem by FP_π.

Since π is given and machines are fixed in certain positions, problem FP_π consists of assigning operations to machines in such a way that operations are compatible with the machines are assigned to, and conditions (1) holds.

Problem 3 [FP_λ.] *Given a feasible loading λ, find a feasible positioning π such that condition (1) holds.*

We denote this problem by FP_λ.

3.3.2 Feasibility given π

If G is a tree, by far the most usual situation, FP_π can be solved in polynomial time with the procedure sketched in Figure 3.3.2. We denote by r the root of G. Quantity $a(i, k)$ will be equal to 1 [0] if there is [there is not] a feasible solution in the subtree of G rooted in i with the condition that operation i is assigned to machine k. Let M_i denote the set of machines able to perform operation i, and P_i, $i \in N$, the set of predecessors of i. It is easy to find a feasible assignment of operations to positions, once $a(i, k)$ are known for each $i \in N$, $k \in M$ by a backward visit of the graph G.

Theorem 2 *If $G = (N, A)$ is a tree , then* FPOP *either finds a solution of FP_π or prove infeasibility in $O(m^2n)$ time units.*

3.3.3 Feasibility given λ

We are now giving some complexity results on FP_λ. Given a feasible loading $\lambda = \{S_1, \ldots, S_m\}$ it is possible to define a *transportation graph* $T = (\bar{M}, R)$, where $\bar{M} \subseteq M$ is the set of m machines selected by loading λ, and $R = \{(h, k) : (i, j) \in A, i \in S_h, j \in S_k\}$. We have the following result:

Theorem 3 *FP_λ is NP-complete even if $T = (\bar{M}, R)$ and $E = (P, C)$ are both trees and the machines are compatible with all the positions.*

It is useful to consider the *condensation* of the graphs $T = (\bar{M}, R)$ and $E = (P, C)$ as defined in [16]. A strongly connected graph is a directed graph where, for each

---

Procedure FPOP (Feasibility Problem: Operations to Positions assignment)

Input: Assembly tree $G = (N, A)$, set of compatible machines M_i for each operation i, π: position of each machine in E.

Output: A feasible assignment, if any, π_i for each operation i.

Comment: We denote by an integer $h = 1, \dots, m$ both the machine and its position.

1. For each leaf $i \in N$ and for $h = 1, \dots, m$, let:

$$a(i,h) := \begin{cases} 1 & \text{if } h \in M_i; \\ 0 & \text{otherwise.} \end{cases} \qquad (3.3)$$

2. Repeat the following until the root r is the only node left in G:
   (a) Consider any node i such that all its predecessors are leaves of G. Then for each $k = 1, \dots, m$ let $a(i,k) = 1$ [0] if $k \in M_i$ and for all $j \in P_i$ there exists a machine h such that $a(j,h) = 1$ and there is a path from h to k in E [otherwise];
   (b) Delete from G all the predecessors of node i.
3. FP_π has a solution if and only if there is a machine $h \in M_r$ such that $a(h,r) = 1$.

FIGURE 3.6. Procedure FPOP

pair of nodes i and j there is a directed path from i to j. We say *strong component* of a graph any maximal strongly connected subgraph. Let $C_1, C_2, \dots C_s$, be the strong components of a graph $G = (N, A)$. The condensation of $G = (N, A)$ is a graph $G^* = (N^*, A^*)$ having the strong components of G as its nodes, and an arc in G^* from node C_i to node C_j whenever there is at least an arc in G from a node of component C_i to a node of component C_j. (See Figure 3.7.) Clearly, a condensed

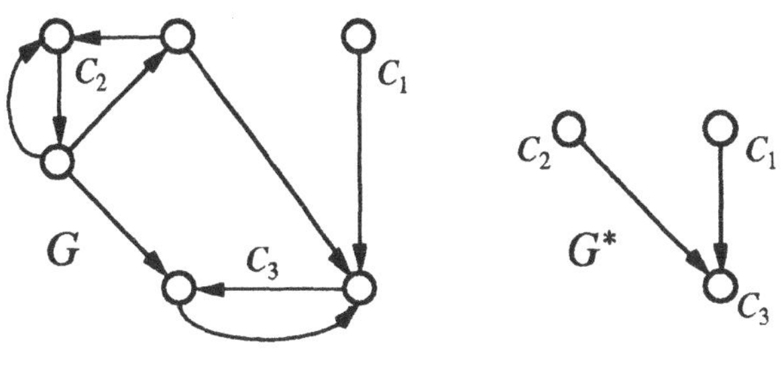

FIGURE 3.7. A graph and its condensation

graph is acyclic. For our purposes, we give each node a weigth equal to the cardinality $|C_i|$ of the corresponding strong component of G.

In particular, we denote by $T^* = (M^*, R^*)$ and $E^* = (P^*, C^*)$, the condensations of graphs T and E, respectively. Moreover, let $v(i)$, $i \in M^*$ and $w(j)$, $j \in P^*$, the node weights in the graphs T^* and E^*, respectively. Obviously the following equalities hold:

$$\sum_{i=1}^{|M^*|} v(i) = \sum_{j=1}^{|P^*|} w(j) = m. \tag{3.4}$$

A straightforward consequence of the definition above is the following necessary condition for the existence of a solution of FP_λ.

Remark 2 *A solution of FP_λ exists only if $|M^*| \geq |P^*|$.*

A simple heuristic for FP_λ is illustrated in Figure 3.8. Clearly, if this heuristic is

---

Procedure **FPMP** (Feasibility Problem: Machines to Positions assignment)

Input: Wheighted graphs $T^* = (M^*, R^*)$ and $E^* = (P^*, C^*)$.

Output: A solution of FP_λ.

1. Number the nodes of $T^* = (M^*, R^*)$ and $E^* = (P^*, C^*)$ according to a topological order (therefore the node in the last position has the maximum index).
2. Repeat the following until either a feasible solution is found or a stop condition is verified:
   (a) Choose the node i of $T^* = (M^*, R^*)$, not yet assigned, with maximum index (if no such a node i exists then a feasible solution has been found) and the node j of $E^* = (P^*, C^*)$, with maximum index and weight $w(j) > 0$.
   (b) If $v(i) > w(j)$ then STOP (no solution is found), else the node i is assigned to position j and $w(j) := w(j) - v(i)$.

FIGURE 3.8. Heuristic **FPMP**

repeated for all the possible topological orders of both T^* and E^*, then we are guaranteed to find a feasible solution, if it exists. Nevertheless in some cases we are guaranteed to find a feasible solution by applying the heuristic to only one order. For example, if both T^* and E^* contain spanning paths, then clearly there exists only one topological order for each graph, and we are done. Another straightforward but important case is when E^* contains a spanning path, and $v(i) = 1$ for each $i \in M^*$ (e.g., if T is acyclic and E^* is a line). In fact, in this case, it is easy to see that, for any topological order induced by T^*, **FPMP** finds a solution of FP_λ, or proves its infeasibility.

On the other hand, if the condition $v(i) = 1$ for each $i \in M^*$ does not hold, the problem falls in the class of NP-complete problems, as shown below.

Theorem 4 *FP_λ is NP-complete, even in the case of T^* tree, E^* path and each machine compatible with each position.*

3.4 Optimization properties

In this section, we deal with the problem of finding one or more optimal decisions for some polynomial cases of Problem 1, being fixed a subset of the decision variables.

3.4.1 COMPLETION TIME MINIMIZATION

In particular we consider Problem 1 in the following polynomially solvable case: (*i*) G is a tree, (*ii*) E contains a spanning path, (*iii*) each machine is compatible with each position, and (*iv*) a connected loading $\lambda = \{S_1, \dots S_m\}$ is given. The objective function is the minimization of the completion time, i.e. the time necessary to complete all the operations. The problem here consists of finding optimal partitioning π, routing ρ and sequencing σ.

Since λ is connected and G is a tree, each subgraph of G induced by $S_h, h = 1, \dots m$ is a tree. Therefore, for each $h = 1, \dots m$ there is a node $r(S_h)$, root of the subtree induced by S_h. If there is a directed path from a node $j \in S_h$ to a node $i \in S_k$, $k \neq h$, then $r(S_h)$ belongs to this path. Observe that, the completion time C_h of operation $r(S_h)$, for each $h = 1, \dots m$, corresponds to the completion time of all the operations assigned to machine h. Let us introduce a release date r_j for each operation j of the subgraph:

$$\begin{array}{ll} r_j = 0 & \text{if } j \text{ has no predecessors} \\ r_j \geq r_i + p_i & \text{if } i \in N' \text{ preecedes } j \\ r_j \geq C_k & \text{if } j \text{ is the successor of machine } k \end{array} \tag{3.5}$$

Here p_i denotes the processing time of operation i. Note that, as soon as the operations assigned to a machine are scheduled, the release dates will change according to (3.5), since new arcs are added to the set A'. Since dummy operations must be scheduled, they will affect the completion time. A lower bound for the values of release dates and completion times can be found in the hypothesis, that no dummy operation exists or they do not affect system behaviour.

Since $E = (P, C)$ contains a spanning path, let first address the two following scenarios. In the first case we consider a *complete acyclic* digraph $E = (P, C)$. In the second case we concern with a *line*, i.e. a digraph $E = (P, C)$ where $C = \{(h, h+1), h = 1, \dots |P| - 1\}$. We will show that any "intermediate" situation can be viewed within these two scenarios.

We suppose the nodes of E are numbered according to the (only) topological order.

Case 1: E is a complete acyclic graph. In this case the problem of minimizing the makespan is trivial: any feasible positioning π is optimal (see Theorem 1 on feasibility). The decision ρ^* introduces no dummy operations; in fact for each pair of position h, k, either there is a directed connection or there is no path connecting them. Once π and ρ has been decided, the only decision left is to schedule operations with given release dates, on each machine, in order to minimize completion times. Hence, an optimal sequencing σ^* is simply obtained by processing first the operations with earliest release dates (ERD rule). The Procedure `R_dates` (described in Figure 3.9) finds σ^* by computing, for each machine, the minimum completion time satisfying equations 3.5. Clearly, the values of completion times equal the release dates of their successors and the makespan is equal to $\max\{C_1, \dots C_m\}$.

Case 2: E is a line. Also in this case we can achieve the same value of the makespan of the previous case. In fact, the simple dynamic programming Procedure `Min_makespan` described in Figure 3.10, finds a positioning π^* in such a way that dummy operations do not affect the completion times of all the other operations.

Clearly the makespan obtained for Case 1, is a lower bound of the optimal solution for any graph E (that contains a spanning path). On the other hand, Procedure `Min_makespan` applies to any graph E, provided that it contains a spanning path, and finds the same makespan value obtained for Case 1.

In conclusion, from the above discussion, the following Theorem 5 holds.

Theorem 5 *If $G = (N, A)$ is a tree, $E = (P, C)$ contains a spanning path and λ is given and connected, then the problem of finding a solution $(\sigma^*, \pi^*, \rho^*)$ minimizing the makespan, is polynomial.*

Procedure `R_dates`

Input: assembly tree $G = (N, A)$ with operations processing times, connected partition λ and corresponding graph $T = (M, R)$.

Output: decision σ, values of minimum completion times C_h, for each subset of operations S_h $(h = 1, \ldots, m)$.

1. Initialize release dates values: $r_i = 0$ for each $i \in N$.
2. Let $L = \{\text{current leaves of } T\}$.
3. If L is empty STOP.
4. For each $h \in L$ do the following:
   (a) Compute C_h by scheduling operations with an ERD rule (decision σ.)
   (b) Let $r_x := C_h$, where $x \in N$ is the successor of S_h.
5. Update T, by deleting all the current leaves: $M := M \setminus L$.
6. Go to 2.

FIGURE 3.9. Procedure `R_dates`

Procedure `Min_makespan`

Input: graphs $T = (M, R)$ (tree) and $E = (P, C)$ (line), values C_h of completion times for each subset of operations assigned to a machine $h \in M$. ($|M| = |P|$.)

Output: feasible positioning π^* minimizing makespan.

1. Assign the root r of T to the last position of the line: $\pi^*(r) := m$.
2. Sort the $|P_r|$ predecessors of the root in increasing order of completion time ($C_1 \leq C_2 \leq \ldots \leq C_{|P_r|}$).
3. For each $h \in P_r$ perform `Min_makespan` for the subtree of T rooted in h (call it ST_h) on the part of the line E from position $2 + \sum_{i=1}^{h-1} |ST_i|$ to position $1 + \sum_{i=1}^{h} |ST_i|$.

FIGURE 3.10. Procedure `Min_makespan`

Theorem 5 shows that makespan cannot be reduced by simply adding arcs to the spanning path.

3.4.2 CYCLE TIME MINIMIZATION

Let us now turn to the cycle time minimization. In a relevant special case, we search for a connected loading λ such that the maximum machine workload is minimized. For this purpose, if G is a tree, each operation is compatible with each machine and each machine is compatible with each position, there exist efficient (polynomial) algorithms finding an optimal loading [7, 6]. Moreover, if E contains a spanning path, it is easy to see that optimal routing and sequencing can be found with arguments similar to those described in the previous section. (On the other hand, NP-completeness of the problem for general λ follows from PARTITION, see [14]).

3.4.3 PART TRANSFER MINIMIZATION

If loading λ is given, E is a path and we search for a positioning π that minimizes the number of part transfer, the problem complexity is different whether T is an acyclic graph or a tree. In particular, in [11] the NP-completeness of the first problem is shown, while in [1] Adolphson and Hu propose a polynomial algorithm for the second problem. A more detailed discussion on the part transfer minimization can be found in [22].

In conclusion, Table 3.2 summarizes the complexity results concerning the optimization problem when E is a line.

Complexity / *Objective*	POLYNOMIAL	NP-COMPLETE
FEASIBILITY	*Given:* π, G tree, *Find:* λ feasible. (See theorem 2 on FP_π.) *Given:* λ, T acyclic, E^* path, *Find:* π feasible. (See Section 3.3.3.)	*Given:* λ, T tree, E tree, *Find:* π feasible. (See theorem 3 on FP_λ.) *Given:* λ, T^* tree, E^* path, *Find:* π feasible. (See theorem 4 on FP_λ.)
CYCLE TIME	*Given:* G tree, E path, no incompatible pairs, *Find:* λ connected, π, ρ, σ optimal. (See [7, 6].)	*Given:* G tree, E path, no incompatible pairs, *Find:* λ optimal. (See [14].)
COMPLETION TIME	*Given:* λ connected, G tree, E containing a spanning path, *Find:* π, ρ, σ optimal. (See theorem 5.)	*Given:* λ, π, ρ, *Find:* σ optimal. (See, for example, [14].)
PART TRANSFERS	*Given:* λ, T tree, E path, *Find:* π optimal. (See [1].)	*Given:* λ, T acyclic, E path, *Find:* π optimal. (See [11].)

TABLE 3.2. Complexity results

3.5 Conclusions

The paper outlines and analyzes a unified framework for designing (or re-designing) the configuration of a given production plant and the corresponding network of material flows. An integrated approach could produce substantial advantages in many practical cases, however, global decision models fall into the format of a large optimization problem, often too difficult to solve for real size applications. In such cases it is possible to follow a decomposition approach, i.e. determine a subset of decisions variables at a time, being the others fixed or considered in a next phase of the solution procedure.

In the first step, we find the value of λ (or π) on the ground of practical considerations. For instance, λ can be obtained in the process-plant integrated design phase by clustering operations on a set of m existing machines (or π can be obtained in the layout design phase on the ground of machine-positions compatibilities).

In the second step, we find the remaining decision π (or λ) on the ground of some optimization or feasibility models (FP_λ or FP_π respectively).

Finally, in the third step, we can find the routing ρ and the scheduling σ in the way most of managers do, or by more sophysticated methods exploiting the particular structure of the problem, like those presented in this paper.

In large scale production (e.g., in car components and electronic units production), most of the operations must be processed by dedicated machines and only few handling operations can be performed by different devices. Moreover, the strong precedence constraints among operations, strongly simplifies the downstream decision process. In the small/medium scale flexible production, machines without tools are already positioned in the plant. In this case the positioning problem is actually a tooling problem, usually with few feasible alternatives. The loading problem is, in this case, the critical point of the production strategy. The integrated framework addressed in this paper helps in finding "good" approaches to these problems enlighting appropriate solution strategies.

Future research work will be devoted to the analysis of relevant applications to validate this sistematic framework for solving decision problems up to now approached on the ground of "practical considerations".

3.6 References

[1] A. Adolphson, T.C. Hu, *Optimal Linear Ordering*, SIAM Journal of Applied Mathematics, 25, (3) 403-423, 1973

[2] A. Agnetis, M. Lucertini, F. Nicolò, *Tool Handling Synchronization in Flexible Manufacturing Cells*,Proceedings of the 1991 IEEE Conference of Robotics and Automation, Sacramento, CA, April 1991

[3] R.G. Askin, C.R. Standridge, *Modeling and analysis of manufacturing systems*, John Wiley &Sons, NY, 1993

[4] I. Baybars, *A survey of exact algorithms for the simple assembly line balancing problem*, Management Science, 32, 909-932, 1986

[5] M.S. Bazara, *Computerized Layout Design: A Branch and Bound Approach*, AIIE Transactions, 7, (4) 432–437, 1975

[6] R.I. Becker, Y. Perl, *A shifting algorithm for tree partitioning with general weighting functions*, Journal of Algorithms, 4, 101-120, 1983

[7] R.I. Becker, Y. Perl, S.R. Schach, *A shifting algorithm for min-max tree partitioning*, J.A.C.M., 29, 58-67, 1982

[8] M. Berrada, K.E. Stecke, *A branch and bound approach for workstations load balancing in flexible manufacturing systems*, Management Science, 32, 1316-1335, 1986

[9] P. Brucker, *Scheduling Algorithms*, Springer-Verlag, Berlin, 1995

[10] R.B. Chase, *Survey of paced assembly lines*, Industrial Engineering, 6, 1974

[11] S. Even, Y. Shiloach, *NP-completeness of several arrangement problems*, Report n. 43, Dept of Computer Science, Technion, Haifa, Israel, 1975

[12] S. Ghosh, R.J. Gagnon, *A comprehensive literature review and analysis of the design, balancing and scheduling of assembly systems*, International Journal of Production Research, 27 (4), 637-670, 1989

[13] A.L. Gutjahr, G.L. Nemhauser, *An algorithm for the line balancing problem*, Management Science, 11 (2), 1964

[14] M.R. Garey , D.S. Johnson, *Computers and intractability*, Freeman, 1979

[15] A.E. Gray, A. Seidman, K.E. Stecke, *Decision models for tool-management in automated manufacturing*, Management Science, 39 (5), 549–567, 1993

[16] F. Harary, *Graph Theory*, Addison-Wesley Publishing Company, Inc., Reading, Massachusetts, 1969

[17] M.M.D. Hassan, *Machine layout problem in modern manufacturing facilities*, International Journal of Production Research, 32 (11), 2559-2584, 1994

[18] W.B. Helgeson, M.E. Salveson, W.W. Smith, *How to balance an assembly line*, Management Report no. 7, New Caraan, Carr Press, Division for Advanced Management, 1954

[19] S.S. Heragu, A. Kusiak, *Machine layout problem in flexible manufacturing systems*, Operations Research, 36 (2), 258-268, 1988

[20] R.V. Johnson, *Optimally balancing large assembly lines with FABLE*, Management Science, 34 (2), 240-253, 1988

[21] M. Lucertini, D. Pacciarelli, A. Pacifici, *Layout constraints in assembly problems*, Proceedings of the 1994 Japan-U.S.A. Symposium on Flexible Automation, Kobe (Japan), 11-18 July 1994

[22] M. Lucertini, D. Pacciarelli, A. Pacifici, *Optimal flow management in flexible assembly systems: the minimal part transfer problem*, Systems Science, 22 (2), 1996

[23] Y. Monden, *Toyota production system*, Industrial Engineering and Management Press, Norcross, GA

[24] K.E. Stecke, J.J. Solberg, *Loading and control policies for a flexible manufacturing system*, International Journal of Production Research, 9 (5), 481-490, 1981

[25] J.A. Ventura, F.F. Chen, Wu Chih-Hang, *Grouping parts and tools in flexible manifacturing systems production planning*, International Journal of Production Research, 28, 1039–1056, 1990

3.7 Appendix: proofs of theorems

3.7.1 Proof of theorem 1.

Necessity is trivial. In fact, by construction, if G' is acyclic then $L = \emptyset$ that is in turn equivalent to condition (1).

As sufficiency is concerned, we shall prove that if λ and π are feasible, i.e., each operation (each machine) is assigned to a compatible machine (position), and condition (1) is satisfied then it is possible to find ρ and σ such that G' is acyclic.

Let $G'' = (N', A')$ denote the graph obtained from G, after the decisions λ and π. If condition (1) is satisfied, $L = \emptyset$, i.e., for each arc $(i, j) \in A$, either $\pi_i = \pi_j$ or there is a path from π_i to π_j in E. Therefore there is always a routing ρ such that G'' is acyclic.

Consider now a topological order on $G'' = (N', A')$ (this always exists, since G'' is acyclic) and number accordingly the nodes of G'. For each pair (i, j) of nodes of N', such that $i < j$, there is no path from j to i. A feasible σ can be simply obtained by sequencing the operations of N' assigned to the same machine, respecting the topological order. This choice cannot induce cycles in G'. This completes the proof. □

3.7.2 Proof of theorem 2.

We will prove the theorem by induction on the depth of the tree. We state by inductive hypothesis that **FPOP** applies for a tree with depth less or equal to k (the depth of a directed tree is the lenght of its longest path in terms of number of arcs) and then we will show how **FPOP** holds for another tree with depth equal to $(k+1)$. It is easy to verify that **FPOP** applies when $k = 0$. We know by theorem 1 that a feasible solution exists if and only if λ is a feasible loading and condition (1) holds. Thus $a(i, k) = 1$ if and only if $k \in M_i$ and for each $(j, i) \in A$, there exists $h \in M$ such that $a(j, h) = 1$ and there exists the path from h to k in E.

If we consider a $(k+1)$-deep tree, the Procedure **FPOP** applies for all the subtrees rooted in j, where $j \in P_r$. Suppose there is a feasible solution for each of these subtrees: we can write $a(r, k) = 1$ if and only if $k \in M_i$ and (1) holds for each $(j, r) \in A$. The last one is the condition expressed by **FPOP**, thus valid also for $(k+1)$-depth tree.

Step (a) of procedure, requires $O(|P_i|m^2)$ time. Therefore the overall computational cost is:

$$\sum_{i=1}^{m} |P_i| m^2 \tag{3.6}$$

i.e., $O(m^2 n)$. The proposition follows. □

3.7.3 Proof of theorem 3.

In order to prove the NP-completeness of FP_λ we make use of the NP-complete decision problem 3-Partition (3P) defined hereafter. (For the definitions and properties of *Max*[*I*], *Lenght*[*I*] and *number problem*, see [14].)

Instance. A set A of $3m$ elements, an integer bound B and an integer size $s(a)$ for each $a \in A$ such that $B/4 < s(a) < B/2$ and $\sum_{a \in A} s(a) = mB$.

Question. Can A be partitioned into m disjoint sets $A_1, A_2, \ldots A_m$ such that, for $1 \leq i \leq m$, $\sum_{a \in A_i} s(a) = B$?

Note that each A_i must necessarily contain exactly three elements of A. 3P is NP-complete in the strong sense [14]. The strong NP-completeness implies the existence of a polynomial function p such that the problem 3P, restricted to only those instances I satisfying $Max[I] \leq p(Lenght[I])$, is still NP-complete.

FP_λ can be defined as follows:

Instance. An acyclic graph $G = (N, A)$ and a feasible loading λ.
Question. Is there a positioning π such that condition (1) is satisfied?

We define a *rooted tree* as a tree in which, for each subtree, the root is a successor of every node of the subtree. In particular, we consider the special case in which the rooted trees are composed by the root connected with several chains.

Let us associate to each instance of 3P the following instance of FP_λ: let $T = (M, R)$ and $E = (P, C)$ be both rooted tree where $T = (M, R)$ $[E = (P, C)]$ is composed by the root connected with $3m$ $[m]$ chains. Associate to each $a \in A$ a chain of $T = (M, R)$, composed by $s(a)$ nodes, while each chain of E is composed by B nodes, as shown in Figure 3.11. It is straightforward to observe that any

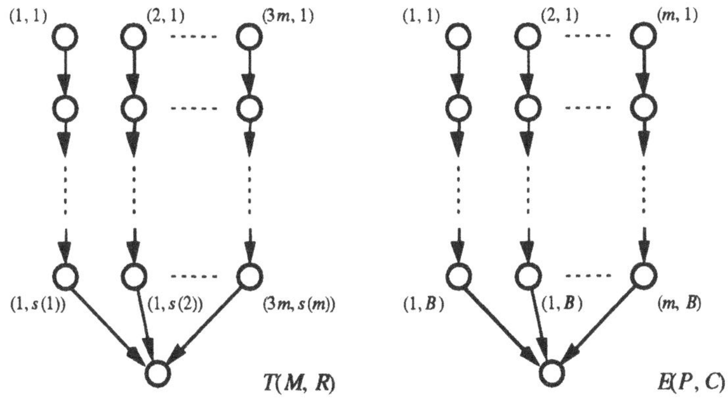

FIGURE 3.11. An instance of FP_λ

instance of 3P is a yes-instance if and only if the corresponding instance of FP_λ is a yes-instance. Moreover, the reduction is polynomial since we can limit ourselves to the case of $Max[I] \leq p(Lenght[I])$. Therefore FP_λ is NP-complete. It is also straightforward to verify that FP_λ is not a *number problem*, therefore FP_λ is NP-complete in the strong sense. □

3.7.4 Proof of Theorem 4.

We will reduce 3-Partition to our problem. Given any instance of 3P, such that B is bounded by a polynomial function in m, we can associate the following instance of FP_λ: let T^* be a star tree with $3m$ leaves, for each $a \in A$ we associate a leaf of T^* with weight $s(a)$. Let E^* be a path with m nodes, each of weight B. Clearly a solution of 3P exists if and only if the corresponding solution of FP_λ exists. Remark that the number of nodes n of the original graph T equals $3mB$, therefore the reduction is polynomial. This completes the proof. □

3.7.5 PROOF OF THEOREM 5.

If E is a line, the Procedures **`R_dates`** and **`Min_makespan`** find a makespan equal to the lower bound obtained in the case of E complete acyclic graph. Therefore the solution is optimal. In any other case, we can obtain the same makespan. As regards the complexity of the procedures, it is easy to compute it in $O(n^2)$ both for **`R_dates`** and **`Min_makespan`**. This completes the proof. □

4

Reactive Scheduling in Real Time Production Control

Erzsébet Szelke[1]
László Monostori[1]

4.1 Reactive operation management - Predictive, reactive and proactive scheduling

Continuous, steady improvement of manufacturing operation management from information technology (IT) perspective is a key requirement to manufacturing enterprises competing under the pressure of changing market demands.

Manufacturing systems often operate in *complex environments rife with uncertainty.* The growing complexity of manufacturing operation management lies in the nonexpected tasks/events, nonlinearities, and a multitude of interactions which arise when attempting to control various activities in dynamic shop floors. This *complexity* and *uncertainty* seriously limit the effectiveness of conventional control and scheduling approaches. *Reactive/proactive systems* of shop floor control and scheduling with *adaptive behaviour* that is ensured by some embedded learning capability and based on real-time monitoring of executed processes, represent the viable alternative.

The broad *goal of manufacturing operation management*, like other resource constrained, multi-agent planning/scheduling problems, is to achieve a *co-ordinated efficient behaviour of manufacturing* in servicing production demands while responding to changes in shop floors in a timely and cost-effective manner [164]. *Operation scheduling*, viewed a major issue of manufacturing operation management, has remained a complex co-ordination task despite the recent trends and efforts towards a structural rationalisation in manufacturing (e.g., through the implementation of group technology and JIT strategies). Quality of the operation scheduling process generally has a profound effect on the overall factory performance. Advance generation of the factory schedules is considered central to co-ordinate manufacturing activities in order to meet some organisational objectives, usually the three-pronged objective of producing parts of higher quality at lower cost within a shorter lead time. The generation of advance (predictive) schedules is also demanded for anticipating potential performance obstacles (e.g., resource contention), and for timely assuring physical preconditions (availability of all the needed resources, presetting of processes) to the execution of scheduled manufacturing activities, in order to minimise the disturbing effects on the overall manufacturing system operation.

---

[1]Computer and Automation Research Institute, Hungarian Academy of Sciences, H-1518 Budapest, P.O. Box 63, Hungary

In industrial practice, however, at least *two factors confound the use of predictive (advance) schedules* as operational guidance [164].

1. Advance schedules are usually generated by scheduling systems that run with static models [140], ignoring important new operating constraints and objectives of live shop operation, thus lacking a close correspondence to the live status of executed processes. Unless basing on a continuous shop floor data acquisition ensured by real-time monitoring of processes execution, these scheduling systems lack a demanded level of fidelity in modelling.

2. They can not cope with the many external and executional uncertainties [172] arising at companies. *External uncertainties* are due to some unexpected production demands raised by the changed market conditions, or to some late deliveries by suppliers etc.; *executional uncertainties* emerge from, e.g. the failed operation or breakdown of machines/equipment, operator absence, under/over estimated value of operation time norms etc., all of which lead to some divergences from plans (advance schedules), and work against efforts to follow predictive schedules as operational guidance in manufacturing processes.

The impacts of the above factors are behind the observed fact that many companies have to re-schedule a high percentage of their planned production [17][68]. According to a recent survey on the state of the art of real-time control (scheduling) issues in FMS (Flexible Manufacturing Systems) by [160], few of the reviewed control and scheduling systems have had the capability of quick re-scheduling needed in practice. Anyway, the performance of manufacturing companies ultimately hinges on their ability to rapidly adapt schedules to the current status of shop floors and the current objectives of shop management. This demand has brought to life the relatively new concept of reactive scheduling [170] which has been coupled with another pragmatic concept, i.e., with proactive scheduling [68], [172], [174]. Both concepts are considered fundamental in the intelligent real-time production control and scheduling of dynamic manufacturing systems, and discussed in detail in the next subsection

The goal of *reactive operation management* in manufacturing is to achieve a *co-ordinated efficient system behaviour* during the execution of manufacturing operations, by responding to changes while keeping the manufacturing processes moving in order to service customer demands in a timely and cost-effective manner [164]. For achieving this goal, it is indispensable for real-time production control and scheduling systems to be enhanced with reactive/proactive scheduling capabilities and also with some embedded learning ability to adapt their own control/ scheduling behaviour, as it is addressed in [172].

4.1.1 Objectives of reactive operation management - Reactive/proactive scheduling

For *reactive operation management* in manufacturing, the main *objective* of achieving a good global manufacturing behaviour (i.e., to globally satisfy some major performance criteria) should simultaneously be balanced with concerns of continuity in operations (because the execution of a schedule sets a large number of interdependent processes in motion), and responsiveness of decision making (to keep the manufacturing processes moving) [164]. *Reactive operation management* in manufacturing is generally considered as a resource constrained, multiagent-type planning/*scheduling decision problem* with the above composite objective which manifests itself in some *major performance measures: (a) economic* (e.g. computation time, cost of disruptions to processes); *(b) operational* (time slot of disruption to execution, and schedule quality with respect to some scheduling criteria, enumerated in the next paragraph).

Scheduling in general is concerned with the allocation of resources over time to perform a collection of tasks [8]. Tasks in discrete manufacturing shop floors can be represented by a number of activities, i.e. operations associated with a set of orders. Temporal precedence relations of operations with their associated resources to produce given parts (orders) are specified by technological process plans. In production operation management, *planning* determines what tasks to perform in a given period, and *scheduling* imposes the time ordering of tasks so as to respect given objectives/constraints [13]. *Job shop scheduling* deals with the allocation of a limited set of resources to a number of activities (operations) associated with a set of jobs (lots of parts, derived from orders) so as to respect given temporal (technological precedence) relations and resource capacity restrictions in order to optimise a set of objectives (scheduling criteria), such as minimise tardiness, minimise work in process (WIP), maximise resource utilisation etc.

Industrial job shop scheduling is a complex problem and difficult to automate [13], [170], [175] for a variety of reasons:

1. It is an NP complete problem with a combinatorially explosive search space at the outset which can then be reduced by the use of various constraints.
2. There may be tight interactions among the scheduling constraints themselves; the scheduling objectives (criteria) are often ill defined, multiple and conflicting (e.g., to minimise WIP inventory while maximising machine/resource utilisation); thus, it is not possible to assess with any precision the impact of scheduling decisions on the global satisfaction of objectives.
3. Definition/evaluation itself of 'what a high quality schedule is' may be fraught with difficulties because of the need to balance conflicting objectives and a trade-off among them. Such trade-offs typically reflect user preferences and the presence of additional domain constraints not captured in the predictive scheduling model [175].
4. Finally, a schedule that is only *predictive* (i.e., it is created by assuming that the world is static and predictable) will be very brittle [140] during its execution time.

With respect to the dynamic and stochastic nature pertaining to live manufacturing environment, it is clear that any effective scheduling system must be able not solely to predict but also to timely react to changes, by properly adapting its control/ scheduling strategies [172], [175] to a changing world. The operation scheduling in live factories cannot be conceived as a static optimisation problem but as a dynamic reactive and proactive process of real-time decision making to achieve the outlined composite goal of management. *Reactive/proactive scheduling* is concerned with the real-time maintenance of a current schedule to make it further executable by re-optimising it during execution time [141]. In a dynamic world, both conflicts and opportunities are offered up for re-optimisation to a reactive/proactive scheduler while the work progresses in the factory floor, the location and measure of resource contention vary in time [150], the performance trends in schedule execution change [174], and thus, the management goals change over time [13]. In the concept of Burke and Prosser [21], as the dynamism of the environment increases (e.g., some scheduling criteria and the problem model may change), it becomes difficult to distinguish between the predictive and reactive behaviour, so the need for real-time systems is felt.

Management of complexity as a result of unexpected changes/disturbances in the manufacturing environment is an issue of high importance in an enterprise organisation [122]. Companies have found different ways to cope with these issues. The most important ones, according to [196] are the following:

- decentralisation of company functions,
- exploitation of experience, creativity and competence of the employees,
- concentration on the core skills of the company.

In the field of manufacturing operation management, over the decomposition of manufacturing systems into smaller units (e.g., manufacturing cells), there are two ways to guide the complexity of operation management problems towards simplicity. These two ways of enhancing the systems' fault tolerance, i.e., to cope with the above enumerated problems are illustrated by Figure 4.1. One way is to enhance the reactivity and proactivity of the scheduling and control systems by sophisticated new control techniques, the other way is to take advantage of decentralised, distributed problem solving, admitted that the two ways may often overlap each other.

According to a recent overview by [170] in the field of reactive scheduling, it is generally conceived as a real-time repair/revision of a complete but execution-time flawed schedule in order to keep it in line with the live status of shop floor processes and events, by re-optimising it upon failure occurrence. *Reactive scheduling* (RS) is generally an *event-driven process* of incrementally repairing/revising the execution time flawed schedule, possibly only its affected parts while re-using its unaffected parts as much as possible. Within an acceptable time span, the RS process should result in an executable schedule that can globally satisfy some major performance criteria such as, e.g., to minimise job tardiness, while maximising resource capacity utilisation and throughput rate, at an acceptable product quality. Reactive scheduling behaviour is assumed as an ability of the scheduling systems for real-time compensating for the effects of a disturbance upon its occurrence in the executed process. Harmonosky and Robohn (1991)[70] define real-time as the immediate response to some event(s) in a manufacturing system such as to a part arrival or machine breakdown in the form of, e.g., re-routing a part, etc. The speed of response, however, may depend not only on the scheduler's response capability compared to system parameters such as magnitude of part processing times, but also on the flexibility and complexity of the controlled system itself.

In addition to a demanded immediate response to unexpected events, it has been recognised increasingly important to ensure a *stable behaviour* of the controlled man-

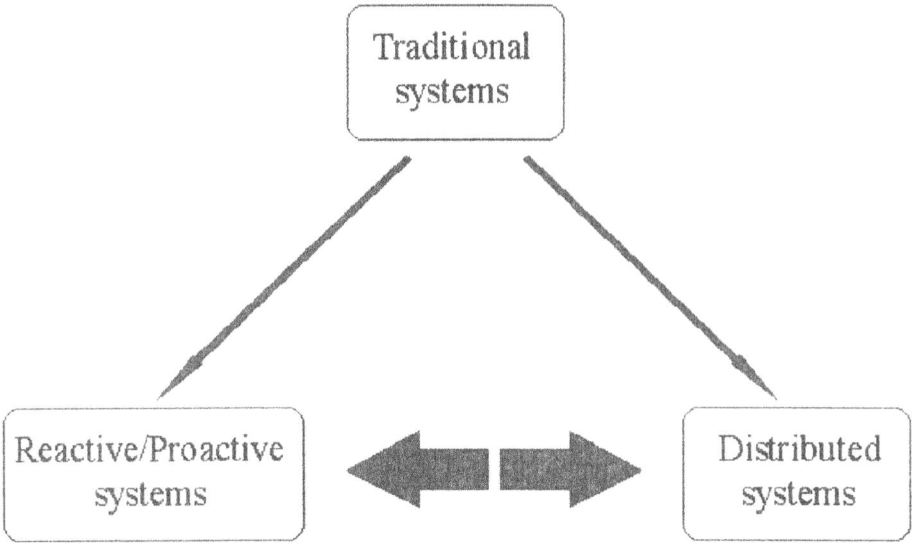

FIGURE 4.1. Two ways for enhancing systems' resistance against changes and disturbances

ufacturing system (i.e., keep the manufacturing processes moving [164]) with respect to the stochastic nature of its operation. The demand for robust schedule from the point of view of execution has led to coupling the reactive ability of an intelligent scheduler with a *proactive capability* in order to avoid time-consuming re-scheduling when possible, and thus, to save stability to executed processes [174]. The proactive capability enables a scheduler to timely prevent some anticipated disturbances or imminent production obstacles, prognosed by the real-time production monitoring function, and helped by a (supposedly correct) stochastic action model for projecting the stochastic outcomes of some prognosed failures. The monitoring function must perform a continual observation of the executed processes with real-time data acquisition, and moreover, a regular sampling of statistics that may reflect major performance trends on the schedule execution in the physical system.

Proactive scheduling (PS) is essentially a *data-driven process* of re-optimising/ adjusting a current schedule to make it fail safe and to prevent some anticipated imminent disturbance/performance degradation during its execution, as viewed by [174]. The term and idea of proactive scheduling have been adopted from [40] and [41], except that they considered the purely reactive approach of re-scheduling as a special case of the proactive approach, and their scheduling and schedule execution system, CERES, was devoted to the real-time operational control of a remotely located automatic telescope (APT at the NASA). However, independently upon the function of the controlled system, if it is a dynamic discrete event system of highly stochastic nature, a proactive scheduler is of any utility to the system operational control, solely if a (correct) stochastic preventive action model (i.e., behavioural model) is available for carrying out a stochastic projection and evaluation of system evolution trajectories, based on the currently observed performance trends and sampled statistics.

According to [41], there are a number of factors that determine whether or not a proactive failure management technique will work. Among these are the nature of the stochastic action "splits". They have hypothesised that if the average stochastic branching factor out of each action is low, and that if the probability distribution over the stochastic outcomes is non-uniform, then there is a good chance that a proactive failure management technique will perform well when compared to conventional schedule repair techniques. The intuition is that when there are few possible failures, and some failures are much more likely than others, then there is useful information in the stochastic action model, and the reasoning performed by proactive scheduling will be of benefit to the system. These and other potential influencing factors should also be considered for further investigation by simulation experiments to elucidate and quantify some of the most important factors for stochastic manufacturing systems to benefit from proactive scheduling.

RS and PS can be conceived as complementary control options usable during a schedule execution in dynamic and stochastic manufacturing environments, for re-optimising a current schedule in order to save/improve its quality and to preserve stability/continuity in the manufacturing activity [174].

4.1.2 MONITORING - A BASIS OF RS/PS IN REAL-TIME PRODUCTION CONTROL

In live manufacturing shop floors, *real-time scheduling decisions* may concern

1. *operation dispatching/scheduling* to generate a new complete advance schedule at a regular daily/shift basis, or to re-optimise an advance schedule, if necessary, by its full re-scheduling from scratch;

2. *RS, i.e., on the spot reaction to unexpected events* (such as, e.g., shortage of resources/supply, machine/equipment breakdown, request for expediting some critical/new orders etc.) to revise a current schedule during its execution (by an RS process) in order to obtain an immediate response to such events in the form of a feasible/executable schedule;

3. *PS, i.e., proactive maintenance* of a current schedule in order to prevent an anticipated failure, and avoid system performance degradation by schedule re-optimisation with respect to the further execution of scheduled manufacturing activities.

While *reactive/proactive scheduling focuses* on the re-establishment of the *schedule feasibility*, improvement of the *acceptability/quality of solution*, considerable pragmatic value lies in *retaining continuity* in the solution produced across iterations (by re-using as much of the advance schedule as possible). It is economically important *to preserve the continuity in domain activity* while making those schedule changes that are necessary to ensure continued feasibility of the schedule and its adherence to overall performance objectives [165]. Both of the two aspects of the schedule maintenance process place a *premium on incremental reactive scheduling* capability.

For achieving better control by RS/PS as control options in a dynamic and stochastic manufacturing environment, the *real-time feedback based monitoring of execution* is a natural demand in real-time production control systems. One of the goals of manufacturing process management is the ability to continuously improve the schedule execution performance, which needs an intermittent adaptation of control strategies to the demands of the current status of executed processes in the shop floor. This ability must be supported by real-time execution monitoring, i.e., an organised way of modelling and measuring processes [187] in order to provide feedback for enabling the above improvement.

Performance trend forecasts, based on a regular sampling of statistics, are also considered as essential feedback information for schedule re-optimisation, and, thus, improving the schedule execution stability. This kind of feedback information is used in [86] for dynamic job-shop scheduling, and in [174] for PS in intelligent manufacturing systems. Real-time feedback based monitoring can solely perform a reliable comparison and evaluation of the scheduled and executed processes, analysis of the problem situation encountered (i.e., a deviation from the planned course of processes), and diagnosis based on a descriptive (behavioural) model of the current system state [146].

In the analysis phase, the system descriptive model [16] is of benefit, solely if based on valid data, and it can correctly diagnose the causes of the problem situation encountered, for increasing the reliability of a following re-scheduling/RS/PS decision making process. To result in feasible schedules (i.e., detailed control plans for schedule execution control and monitoring, based on a valid model of the current system state), monitoring should trace state changes of the evolving shop floor processes, controlled for both detecting unexpected events and ensuring that the control actions for schedule execution are successful. In the phase of diagnosis, faults as deviations from the (advance) detailed schedule/control plan are detected by their impact on the control plan (i.e., the resource availability conflicts/contingencies impacting the schedule feasibility), in order to lead to the adequate control measures (i.e. full rescheduling/RS/PS) pertinent to the occurred fault situation. The control structure for schedule execution, together with the interrelation of the involved control activities treated above, are illustrated by Figure 4.2, related to both a physical and a simulated (virtual) manufacturing environment.

Similarly to propositions by [146], however with an extended, company-wide scope of monitoring and diagnosis, [195] propose their inclusion as new components into production planning (scheduling) and control. Both groups of authors emphasise in their discussions that the representations for the control plan, the control effects and the system state must capture the evolving/temporal nature of manufacturing processes very clearly.

All the real-time scheduling decisions referred to above should be based on a real-time monitoring of the manufacturing processes, including a continual shop floor data-acquisition for tracing any state changing events (such as, e.g., arrival of a new order, start or finish-time of an operation, availability or unavailability of a resource, etc.) in the system. At the stage of real-time control in manufacturing, where the execution of scheduled operations is supervised, the *task of monitoring* is to recognise whether the real performance of the system deviates from its planned and expected behaviour. This routine inspection task [184] can be executed by using *data- and function-based strategies* [188].

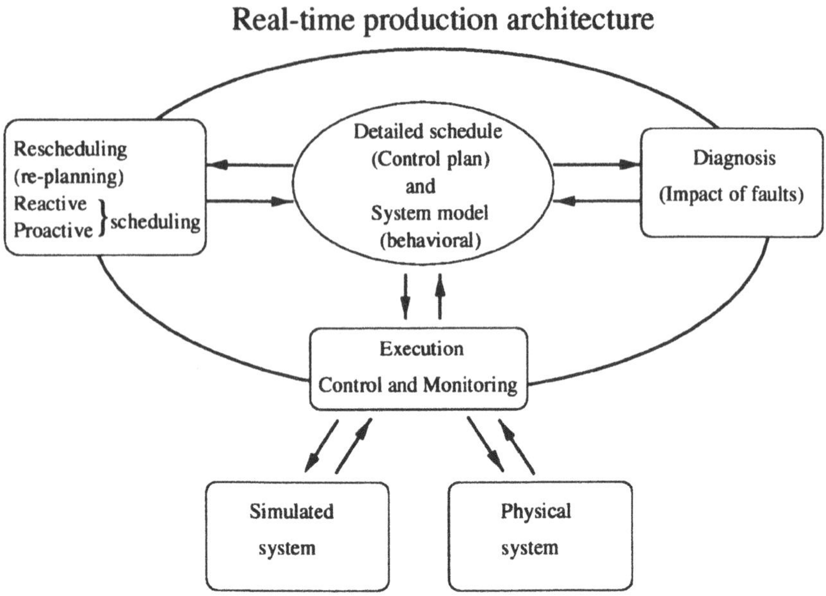

FIGURE 4.2. Real-time production architecture

The clue of *data-driven monitoring* (usable for PS) lies in establishing a correspondence between a particular data pattern (a mixture of sensor and status signals, and projected values, all reflecting forthcoming state change events) and one or more modes of failure. It has to be decided whether one or more of the *a priori* defined symptoms of possible malfunctions/disturbances can be recognised. The quality of the data available to identify symptoms (the data might be noisy, unreliable and incomplete) makes the handling of uncertainties inevitable (e.g. belief value enriched cause-effect networks [188]). The bottom up approach implies that a large amount of data should be processed in real-time, thus, fast rule-of-thumb methods should be applied in order to guarantee real-timeness.

In the heart of any *function-based monitoring* system, there is a hidden, implicitly encoded goal-structure, the primary goals of which are to maintain security, to guarantee system/process survival etc. Of lower priority is the requirement/goal of accommodating the production to a changed environment. However, the goals of specific monitoring activities should be recorded explicitly, in order to be able to evaluate, compare goals, and switch (if need arises) to another monitoring mode. In real-time control systems such a capability is of crucial interest. This inherently top-down approach implies an explicit, hierarchical structure of monitoring goals, which can be established on the basis of the functional properties of the monitored manufacturing system. Taking this approach, a hierarchy of various system functions has to be defined. The performance of these functions can be scrutinised from time to time in the light of the new incoming observations. If the *a priori* selected key functions are performed in an acceptable way, then it is reasonable to assume that the overall system also works properly. Looking at the example of a manufacturing cell, high-level key functions can be defined so that they globally take care of the flows of materials, energy and information in the cell, between the outer world and the cell, and finally check the co-function of cell operations. At the next lower level, more elementary functions may check whether workpieces, tools etc. are correctly dealt with.

The monitoring mechanism can be robust and reliable only if it is sensible enough and, at the same time, has the capability of inspecting certain aspects of the controlled system operation. Thus the event based and the function oriented monitoring processes should run concurrently. Paths of information exchange have to be defined in order to allow the data-driven process to interrupt the top-down checking of key system functions, whenever the sign of a severe malfunction/disturbance is detected by one of the *ad hoc* techniques. The required quick transition between the monitoring processes with different perspectives can be facilitated by e.g. blackboards. For instance, it is indispensable for reactive/proactive schedulers to make decisions based on the real-time monitoring of executed activities from the perspective of both the orders and the resources [131], [163]. Real-time monitoring processes with different perspectives can increase the transparency and reliability of decision making in real-time production control. The different perspectives can provide immediate feedback to the scheduler and achieve better control by, e.g., dynamically *focusing attention* between the *resource* and the *order perspective of scheduling*, as being implemented in some systems (such as ReDs [68], OPIS [131], BOSS [80], and MICRO-BOSS [150].

The monitoring perspectives can further be enriched, if one consider reactive scheduling as a *decision making process of distributed intelligence* in a computer networked factory environment which presumes *direct interaction/ on-line communication* with different 'holons' (i.e., autonomous, intelligent, flexible, distributed, co-operative agents) of the controlled manufacturing environment [122]. As plants generally have different "decision areas", *real-time distributed scheduling systems* like ReDs [68] and DAS [140] as new architectural frameworks have received more attention (see them detailed in 4.1). A distributed architecture can best serve the need for *providing multiple perspectives on a real-time basis* and evaluating efficiency against multiple objectives [68].

Diagnosis to a (prognosed) failure of schedule execution can be supported by an appropriate structural and behavioural model of the manufacturing system. Based on this model of the current system state, expectations on the internal operation of the system or its components can be derived, which, in turn, can be compared with the observed behaviour of the live system. Symptoms, i.e., discrepancies between the expected and observed performance, indicate malfunctions in the live system operation, provided of course that the model is correct. Beside the models for behaviour prediction, heuristic rules can be used [184] to reduce the search space for failure causes and find plausible symptom-cause connections, instead of constructing causal paths. These rules assign symptoms to faults, and are based on empirical association.

In an integrated approach to real-time monitoring and reactive scheduling at [77], for instance, the monitoring is accomplished via behavioural models that characterise the class of admissible system trajectories (i.e., the valid system behaviour). Any trajectory which does not satisfy the behavioural model specifications is considered a fault. This approach uses the on-line observations to construct recursively a representation/encoding of the set of admissible system trajectories that could have generated the current set of observations. They call this encoding of the valid trajectories the evolution graph. The behavioural models characterise the dynamics of the computer controlled process, in the above example: large scale distributed production processes in a steel mill, considered as the interaction of a collection of process variables which evolve dynamically in sets of discrete operating regimes. Behavioural model based monitoring [187] can also be used in manufacturing systems with diverse processes where the pattern based monitoring techniques prove to be inappropriate [77] because it would be difficult to establish the patterns due to variations in facilities/processes.

For proactive scheduling, in addition to the real-time shop floor data acquisition on the system state changes during the schedule execution process, the real-time monitoring should also include [172], [174] a *regular* (i.e., discrete time span) *sampling of collected statistics* reflecting the major performance trends of the evolving execution process, thus providing a reliable basis to the stochastic projection and evaluation on the system evolution. In live manufacturing environment of dynamic and stochastic nature, statistics should continually be collected (by some suitable spreadsheet packages) on the moving averages/standard deviations related to performance measures such as the tardiness of jobs, machine/resource utilisation, the work remained on jobs, throughput rate, WIP levels etc., pertinent to the effectiveness of executing the scheduled activities.

By using the above statistics as patterns for situation recognition, a current assessment of the major performance trends of execution can reflect the goal and guide the demanded direction of a subsequent PS action as control measure to adjust the current schedule (considered as nominal reference to any changes) during its execution. For instance, based on resource usage samples, resource usage distributions can be calculated and bottleneck regions identified [150]. Beside the bottleneck patterns reflecting the resource congestion areas, projections can also be made on processes evolution by using the tardiness patterns of jobs, the breakdown patterns of machines etc., as proposed by, e.g., [68], [174].

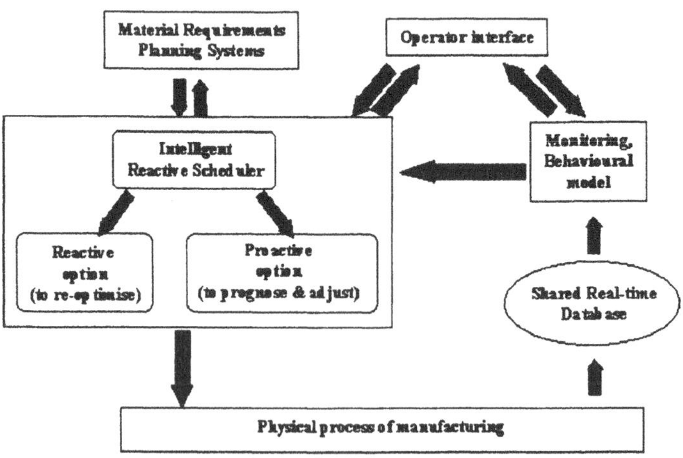

FIGURE 4.3. A control system architecture with RS/PS

A control system structure that may be considered generic to manufacturing environments, by supposing input in the form of draft schedules/plans from an MRP II system (from plant level) is illustrated by Figure 4.3 (after [77]), and with indicating the interrelations of the input planning system, the intelligent reactive scheduler (of RS/PS options), the physical processes, and the real-time monitoring and diagnosis functions based on the system behavioural model(s).

4.1.3 RS PROBLEM COMPLEXITY - IT REQUIREMENTS AGAINST SOLUTION APPROACHES

Over the known complexity of operation scheduling problems as they occur in live industrial environment, *reactive/proactive scheduling* has some *additional complexity* and has been difficult to automate [21], [170] for a variety of reasons:

1. Reactive scheduling, like the *resource constrained scheduling problems* in general, is *combinatorially complex, NP hard problem*, thus computationally unfeasible to be solved by the sole use of conventional Operations Research approaches. AI based or hybrid techniques using domain specific heuristics are necessary to guide the search and to provide satisfying good solutions timely. Hence, it is indispensable for the AI based or hybrid techniques/approaches to the problem to reduce the combinatorially explosive search space of the problem, given at the outset, to a search space of feasible solutions by the application of the many constraints. This demand gave a high *importance of constraint satisfaction techniques* [56], [59], [117], [58] used in various solution approaches to RS problems.

2. There may be *tight interactions among* the scheduling *constraints* themselves; the scheduling objectives (criteria) are often *ill defined, multiple and often conflicting* (e.g., to minimise WIP inventory while maximise machine/resource utilisation); thus, it is not possible to assess with any precision the impact of scheduling decisions on the global satisfaction of objectives. Definition/ evaluation itself of 'what a high quality schedule is' may be fraught with difficulties because of the need to balance *conflicting objectives and a trade-off among them* [13], [175]. Such trade-offs typically reflect user preferences and the presence of additional domain constraints not captured in the predictive scheduling model [175], [171].

   Like in real time operating AI based systems in general [6], the many time dependent data such as e.g. the *time variable constraints and objectives* in reactive/proactive scheduling concerned problem solving result in some additional complexity. It requires a time stamping of events/variable values, the use of truth maintenance, non-monotonic and temporal reasoning techniques for reactive schedulers from IT aspects.

3. Provision of a *feasible/executable solution near real time* so to keep its validity by the time it has been computed and enacted requires a real-timeness/ responsiveness of the scheduler. The utility of a response in the term above [74] often demands trade-offs between the schedule quality and response time efficiency

[101], [175]. Real-timeness demands various methods of representing and reasoning with temporal information, fast situation recognition/assessment, handling temporal logic. Whilst control systems should generally be synchronous and state based (for predictability), reactive schedulers must be capable of responding to asynchronous as well as periodic inputs. In order to reliably and timely control and guide the execution of manufacturing operations, reactive schedulers must have a predictable time performance for providing an acceptable output.

To meet the requirement for an *increased responsiveness and flexibility of decision making* in RS problem situations, agent architectures (with structure and control) combining the complementary advantages of reactive and search-based architectures [118] are required. Reactive systems provide the advantage of quick response, while search-based schedulers offer the flexibility advantage of broad scope for handling a more diverse range of unanticipated system states.

4. For *handling uncertain/incomplete information/knowledge*, mostly available on the controlled system status, valid goals, constraints etc., needs special AI techniques, namely pattern recognition and situation assessment techniques, neuro-fuzzy techniques combining fuzzy logic and neural networks etc. [120]. Combined use of the above techniques is anticipated for proactive scheduling by [174]. For handling uncertainty in job shop scheduling, fuzzy set methodologies are proposed by, e.g., [44], [207], [91].

5. To *reduce the problem complexity* and handle uncertainty are both indispensable for ensuring *the tractability* of reactive scheduling from an IT perspective *in distributed factory environment.* For coping with complexity and ensuring consistency of shop floor scheduling decisions, *distributed problem solving* (DPS) and joint *decomposition techniques* should be taken into consideration, and further complemented by some *co-ordination techniques*, used by e.g., [140], [176], [144].

6. Reactive schedulers *must interact/on-line communicate with their environment*, given as *one major concern of reactive systems* [139]. Live factory floors generally represent a distributed control/ scheduling decision environment. DAI based multiagent reactive schedulers must be specified in terms of their communication with their environment (i.e. with the human and other decision making agents sharing the scheduling task in the manufacturing shop floor). Communication via networks demands an increased activity integration/co-ordination of different agents' activities [127], and real-timeness of information exchange through the basing of decisions on common databases with consistent semantic interface. In live industrial plants there are generally different "decision areas", thus *real-time distributed reactive schedulers* like ReDs [68] and DAS [21] use some special *architectural frameworks*. Such a distributed architecture can best serve the need for *providing multiple perspectives of scheduling* (such as order and resource perspectives) *on a real-time basis* and evaluating efficiency against multiple objectives [68].

7. *Interfacing with the controlled plant, the human, and existing software* of a reactive scheduler demands to fulfil different IT concerns. One of them is that human-system interaction needs a thorough user interface design [171] that allows for inclusion of human preferences in decision alternatives. The design has an impact on the control architecture/mechanism of triggering the problem-solving agents, and providing the user with the relevant segment of information in a timely manner according to the user needs. An intelligent interface design [168] implies as identified the differentiated needs of information users (operators, plant managers, system specialists or knowledge engineers etc.), and is capable of a kind of user cognitive modelling.

   A recent overview series on shop floor problems [133] pointed out that shop operators are particularly weak in *performance evaluation* on scheduled/executed processes. To enhance their responsive capabilities, the intelligent user interface should provide them with some built-in mechanisms that facilitate reactive decision making. *Performance measures* either economic (from shop management) or operational (from shop operator/supervisor) are most often intended to trigger human action. Thus, the performance evaluation against the different measures can be conceived as an integrating link between human and the shop (CIM) environment with its information processes.

8. Finally, a truly reactive scheduler *must adapt its own behaviour* (control/ scheduling strategies) in response to changes in the controlled system status/valid goals. It is best facilitated by *learning from examples.* For this purpose, a case based learning (CBL) approach is used in [175] and another CBR/L by [171]. Inductive learning is preferred by Kerr and Kibira (1995) [90], and learning by neural networks using fuzzy logic is proposed by [174]. Learning from experiments in a simulation testbed is visualised by Figure 4.4, as it was originally outlined and illustrated in [30], a repository of situation/goal dependent strategies for reactive and proactive scheduling can be established to improve the scheduler's problem solving efficiency [121].

4.2 Models of Reactive and Proactive Scheduling Problems

Manufacturing systems from modelling perspective are considered as *dynamic discrete event systems (DEDS)* [191], [52], [76], [181]. For constructing valid models of manufacturing processes by the design and simulation of the real-time discrete control/ scheduling logic, the models should represent the discrete event evolution of the system and its admissible trajectories. Discrete event systems can be further dichotomised [12] in many ways: sequential systems versus concurrent ones, deterministic versus nondeterministic systems, synchronous versus asynchronous systems, etc. However, a more important classification is *reactive versus transformational* systems [69]. A transformational system is characterised as a blackbox in which every input is transformed into an output, assumed to be activated only when all inputs are available. *Reactive systems* on the other hand constantly react to input signals.

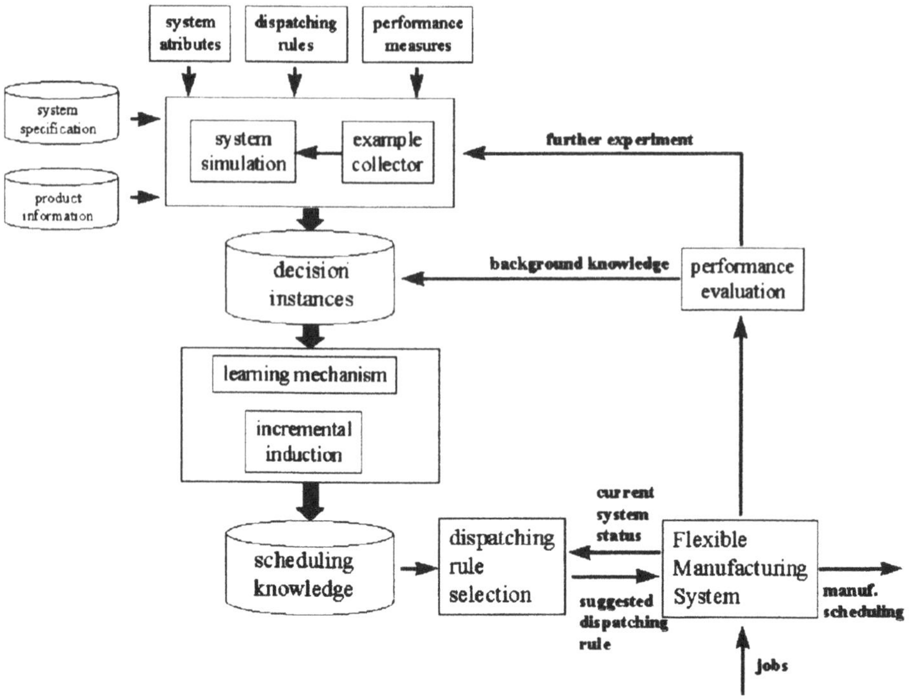

FIGURE 4.4. Architecture of a learning-based dynamic scheduling system(Chiuc and Yih, 1995)

The system components are active constantly, and every input change causes a state change in the system. Most of the systems in practice are reactive, however, due to modelling difficulties, some are looked at as transformational. *Modelling* reactive systems is *inherently difficult*: there is no simple transformation that can represent the system. The reactive system is usually hierarchical, and its components are concurrently and asynchronously active, like operating manufacturing systems [12]. The communication among the components is a crucial part of the model, and the variety of states that the system can have is very large.

Reactive systems are systems that can best be described by *behavioural models* [139] because they cannot be adequately represented by *relational (structural)* or *functional* models/views, conventionally used for system description. The relational (structural) view represents the structure of the system, its components, and links among the components, regards programs as functions from an initial state to a terminal state. Typically, the main role of reactive systems is to maintain an interaction with their environment, and therefore must be described (and specified) in term of their on-going behaviour. Some of the interaction may also include a final result, but other reactive systems such as process control systems, RS systems are not supposed to terminate. Thus, *for RS systems* that are embedded and operating real-time in shop floor processes, the *behavioural view*, representing the various states the system goes through and the flow of process, is inescapable in addition to the relational/structural and the functional view of modelling. Every concurrent system (like real-time distributed scheduling systems), regardless of whether its role is computational (automatically can produce the final result) or interactive, must be studied by behavioural means [139]. This is because each individual module in a concurrent system is a reactive sub-system, interacting with its own environment which consists of the other modules.

The various approaches to RS (see a current overview in [170] are based on different paradigms. They often use the knowledge-based paradigm hybrid with simulation oriented or graphical modelling frameworks/tools for modelling/analysing the operating manufacturing system behaviour, therefore, supporting solution of RS problems that mostly emerge in distributed factory environment with concurrent processes.

System modelling methods include a wide inter-disciplinary variety of methods, including graphical techniques, simulation, formal languages, AI-based approaches, DAI frameworks, etc. The following review, with a special emphasise on real-time control perspectives of manufacturing management, is by no means comprehensive, and is intended to highlight the major modelling approaches of reactive scheduling systems. Mathematical and analytical techniques like Markov chains and queuing networks cannot describe the causal relation of uncertain events explicitly in their complex static models because of their unrealistic mathematical assumptions [72], and thus, not aimed for reactive systems and excluded from the following review of modelling techniques.

4.2.1 GRAPHICAL MODELLING TECHNIQUES

Graphical specification is natural and easier means of expressing system functionalities [106] because it carries the inherent advantage of problem visualisation.

Petri Nets

One of the most attractive methods for reactive systems modelling is Petri nets [138]. Petri net (PN) is a graphical and analytical methodology for system modelling and analysis, and a large body of modelling work has been conducted using this approach (e.g., [43]). Petri nets are well-known techniques for modelling the dynamic behaviour of distributed systems (like Flexible Manufacturing Systems, FMSs) in which concurrency and synchronisation play a major role [127]. Originally they were invented for modelling those aspects of system behaviour which can be expressed in terms of causality and choice. An extensive amount of research has been performed in modelling Flexible Manufacturing Systems (FMS) or manufacturing workcells [23], designing control logic for FMS [199], and analysing performance and decision outcomes [187], also in distributed environment [106], by using Petri nets.

Simple Petri nets as a reactive modelling tool, however, suffer from some limitations among which are: limited expressive power, and lack of explicit time information, the latter is vital in real-time control. In order to overcome some of these limitations, many extensions to Petri net models have been proposed for use in practical modelling [189]. Some include time explicitly (timed Petri nets [18]), others allow greater variety of tokens (coloured Petri nets [83], [87]), (Process-translatable Petri nets [19]), or allow to explicitly describe the causal relation of uncertain events (stochastic Petri nets [72]).

The *modified PNs* above proved to be more powerful techniques [1], [97] for rapid prototyping of real-time control systems than queuing networks or ordinary PN. They might also be taken into account for configuring, analysing, and evaluating different solution alternatives in case of RS problems, conceptualised as stochastic control problems and modelled as stochastic Petri nets, e.g., at [72]. Some of the extended models have been used for communication protocols verification in DAS [182]. For experimenting with generic FMS real-time control and scheduling policies embedded in a rule based (Prolog) expert system, [28] use *coloured Petri net* as a powerful tool for modelling and analysing the discrete event-driven system of a generic operating FMS. A survey on further Petri net applications in modelling controls for automated manifacturing systems is available in [42].

Scheduling graph (SG)

It is considered as a direct means of RS and an alternative representation structure to the known schedule Gantt chart has recently been reported by [200]. It represents the interrelations between various scheduled jobs, between operations on the same scheduled job, and between scheduled jobs and machines. The SG together with the new concepts of time effect and relationship effect of a schedule compression, as introduced by [201], provides means of : (1) identifying those operations that require revision in a current schedule; (2) revising the identified/affected operations (via a partial change in the scheduling graph structure), and (3) updating the start/finish times of the revised operations.

4.2.2 SIMULATION

Simulation modelling tools can be divided into two categories: simulation languages, and system simulators . *(1) Different simulation languages* (e.g. SLAM II, SIMAN, CML) tend to be flexible and powerful tools, sometimes linked with animation like SIMAN/CINEMA. *(2) System simulators* (such as e.g. SimFactory, SimView, SEE-WHY, WITNESS, ARENA) tend to be object-oriented/menu-driven. System simulators are easier to master and are user-friendly. Additionally, the implementation of graphics, animation and built-in statistics' analysis makes them very powerful tools. Both groups are available for multi-faceted modelling and discrete-event or object-oriented simulation [209], [211], and widely used for factory floor modelling, simulation, and analysis in logistics. However, merely some of them are used for experimenting with strategies of RS problem solving in a real-time environment. These, e.g. the one reported in [65], can support to analyse problem situations, and to test the feasibility/evaluate the performance of various alternative control and scheduling scenarios/strategies. In RS, hybrid knowledge-based and simulation environments [212] combine the power of artificial intelligence and system modelling concepts. This facilitates simulation modelling of knowledge-based control/RS strategies for flexible and autonomous intelligent manufacturing systems [203], [112]. A hybrid simulation framework and genetic algorithm are used by e.g. [195] for decentralised production scheduling/re-scheduling of assembly systems. However, in hybrid environments, interruptible and distributed expert systems can also be defined/studied as components of simulation models [212].

SIMAN [136] as a modelling framework is based on the system theoretic concepts developed by Zeigler and Oren [209] [129] [130]. It allows for a distinction between the system model and the experimental frame to model and analyse the dynamic aspects of live manufacturing systems. Given the system model and its experimental frame, the *SIMAN* simulation program generates an output file recording the system state transitions as they occur in simulated time, hence reflecting reactive system behaviour.

ARENA [67] is an object oriented modelling/simulation environment developed from *SIMAN/CINEMA* for template-based simulation and animation of various systems, e.g. operating flexible manufacturing systems (FMSs). *ARENA*, issued by the System Modelling Corporation Inc., has been provided with a general collection of modules that cover most of the *SIMAN* functionality (to study issues of resource utilisation, material handling control logic etc.). *ARENA* allows for grouping the shop operators and resources that share similar tasks into sets, thus for rapid modelling of teams sharing manufacturing tasks management. It provides the ability to conduct multiple runs in sequence, while parameters are modified between runs that enables to automate the execution of several scenarios. *ARENA* offers power in modelling decision logic. The above features make it a candidate object-oriented environment for selecting a viable alternative from different RS scenarios (e.g. with shortest path of transport, shorter idle time on a next destination machine etc.).

4.2.3 Concurrent Modelling Language (CML)

CML declaratively addresses discrete process modelling, simulation and control of *reactive systems*. Modern manufacturing systems can be viewed as distributed concurrent discrete processes, using extensive communication internally and with the environment. The concurrency and reactive nature of the system make it difficult to model and analyse correctly. CML, as described in [12], is a high level and concurrent modelling language for the simulation modelling and control of manufacturing processes, particularly useful for simulation and analysis of the manufacturing system operation. The CML language is based on flat concurrent Prolog, which is a declarative concurrent logic programming language, therefore, highly expressive and theoretically sound. The CML uses the FCP mechanisms such as atomic unification and guarded clauses. CML is hierarchical and has two main building blocks: activities and communication channels. Activities can be compound, composed of other activities, or primitive at the lower level. Communication channels are used to connect the activities and send messages to activate or acknowledge the activities, therefore preserving the reactive nature of the system in the model. Each primitive activity is defined in terms of its activation conditions and action set. When the activation conditions are met, the set of actions is performed. Those actions include internal state change, message transmission, and resource acquisition.

CML language is reactive in the sense [12] that system components are represented as computational processes that alive concurrently, each reacting to its inputs. During the model operation all the activities are executed in parallel, thus generating a reactive model of a system. CML is a good modelling tool in the sense that it can capture all the three views of the reactive system: *(1) Structural view*, by representing the structure of the system, its components, and the links among the components;

(2) Functional view, reflecting the functionality of the system as a whole; *(3) Behavioural view*, representing the various states the system goes through, and the flow of the process.

Due to the expressive power of the CML language, it offers a variety of analysis methods for syntactic, static, dynamic and hybrid analysis. Syntactic analysis is used for debugging and initial model verification. *Dynamic analysis* is used for performance evaluation and behavioural verification. Such analysis includes *system simulation* performed by executing the CML (FCP) program, emulating the asynchronous operation of the system. Being a reactive model, the various activities thus generate output and communicate with each other, causing state changes and further interaction. When an external input is required, it is supplied by the user that is represented as another activity in the model. Simulation can be performed without specifying activity durations or with such time consideration. Timeless simulation can be performed by CML, which shows the chain reaction of the various activities, the sequence of activity triggering, and information flow used to trigger the activities. The dependency chain of activities or messages (which message triggers which) is useful for system diagnostics and error recovery. If an anomaly is detected during the simulation run, the chain of activities can be extracted thus enabling forward recovery (relevant future activities are detected and altered to compensate for the deviation), or backward recovery in which one of the early activities in the chain is modified to prevent reoccurrence of the error. Such a simulation can also be used to investigate alternative ways to activate a specific operation. By simulating the system and activating various activities, alternative sequences leading to the goal state can be discovered.

A more sophisticated dynamic analysis is performed using controlled execution of the model, in which case a meta-interpreter is used to control the execution of the program. This mode allows step-by-step execution, interrupting the model during simulation for input addition or modification, and using debugging or inspection facilities. The meta-interpreter can execute the model interactively, involving the user as one of the model activities.

Hybrid analysis (allowed due to a special interpreter of CML) can observe the dynamic behaviour of the model (by running the program), but for a general group of inputs. This analysis, therefore, combines some of the generality of the static analysis with the detail of the dynamic one. This analysis allows the model to operate on a group of inputs. The advantage is that all possible outcomes of the input group are investigated, in comparison to a specific branch chosen for one instance of inputs (as typical simulation does). This analysis is found to be especially attractive for models of manufacturing systems whose input are parts families. Thus the system can be analysed in one run for the entire input family, showing longest machining path and possible undesired states.

4.2.4 Graph theoretic modelling used for RS as a Constraint Satisfaction (CS) problem

Factory scheduling in general involves the assignment of start times and resources to a set of operations, supposing that each operation belongs to a job (order). Operations of the same job are subject to *precedence constraints* specified by a process plan. Additionally, no two operations are allowed to use the same resource at the same time that results in *resource capacity constraints*. Each job/order has a release date and a latest acceptable completion date (which may be later than the due date), both are given as *temporal constraints*. In addition, each operation may require one/several resources, for each of which there may be several alternatives.

In live factory environment, the major constraints concerning operation scheduling may be grouped as: *(1) Global preference constraints* that are rather global goals, not necessarily and one-by-one satisfied (i.e., some primary concerns of factory management on taking into account customer importance, maintaining shop stability, minimising work-in-process (WIP) time, etc.). *(2) Local preference constraints* that express the preferred choices from a set of alternative scenarios to scheduling decisions in operation dispatching; these may include preferences for reducing set-up time and for meeting a special resource (e.g. a high precision machine to execute an operation). *(3) Precedence constraints* that restrict the sequence of operations in processing a job/order through the factory.

In addition to the conflict situations between constraints satisfaction and resource availability, typical to many RS problems, there are also problems among the constraints themselves that are mostly interrelated. For example, in order to satisfy the due-date constraints, it is good to release jobs/orders early into the shop to have some slack times. However, this is likely to cause increased WIP time. Similarly to conflict situations between constraints satisfaction and resource availability, it is difficult to predict these problems during predictive scheduling. Constraints bound the solution search space. During the conflict resolution, a dynamic creation and propagation of constraints is accomplished [59]. The propagation of hard constraints (e.g. completion date of customer orders) and relaxation of some soft constraints (e.g. local preferences on set-up times) can manipulate the search space. Typical example is ISIS [59] that uses constraints to bound, guide, and analyse the search.

RS problems can generally be considered as Constraint Satisfaction (CS) problems [183], [58] , [140], [117]. The underlying basis for this approach is that in most constraint satisfaction situations more than one position in the schedule can satisfy a constraint, or typically, there is a range in which a constraint can remain satisfied. Within the satisfaction range, the scheduled operations all can be manipulated (e.g., exchanged [15], [36]) in order to try to satisfy other constraints by exploiting new opportunities.

Constraint satisfaction techniques approach the RS problem by constructing a *constraint graph*. The *nodes* in this *graph theoretic model* are variables with discrete domains, and the arcs are (unary/binary/n-ary) constraints among the values that may be assigned to the variables. Problem solving is performed by sequentially choosing a variable and assigning to it a value that satisfies all constraints incident upon it. The important insight that has been drawn from CS research by, e.g., [58] is that manipulating the constraint graph, the ordering of variables and values can be optimised. That is, the constraint graph model provides a structure for the problem with respect to its solution (search) space. Features of this problem topology are the types of variables and constraints, and their associated propagation algorithms. Fox et al.[58] use so-called 'texture measures' like constraint tightness and reliance, variable value goodness etc. for a more refined description of the problem space state. Their use allows a search strategy to focus attention by a more principled way during solution. Problem objective function may rate in this context states that satisfy their constraints.

4.2.5 Neural Network Models

These models use artificial neural networks (ANNs) which are described by Kohonen [94] as 'massively parallel interconnected networks of simple, usually adaptive elements and their hierarchical organisations which are intended to interact with the objects of the real world in the same way as biological nervous systems do'. In general, a neural network consists of two major parts: the individual processing elements (PEs) and the network of interconnections among them. A neural network may have many PEs in multiple layer topology, in which each PE can operate in parallel with the rest. Knowledge in neural networks is stored in connection weights biases. In this structure, the knowledge is usually represented in a totally distributed fashion which makes ANNs fail-safe (i.e., if some PEs or connections are damaged, the performance of neural networks will degrade gracefully rather than cease to operate completely). Many neural network models are capable of learning by dynamically modifying their structure. ANNs, being inherently provided with adapting abilities by means of efficient knowledge acquisition and embedding [120], are appropriate for empirical learning. There are a number of supervised, un-supervised and self-supervised training and learning schemes that adapt the connection weights to recognise new patterns. These models can meet demands for shop floor control and scheduling due to their real-time capability and learning ability [143] [120].

ANNs exhibit many interesting features such as memory association in which the stored knowledge is used as a key to retrieve or recall the rest of the contents in the memory of PEs. This important feature not only enables neural networks to process incomplete data, but to handle noisy data as well. Another characteristic of neural networks is related to conflict resolution. The networks behave such that all PEs co-operate together to arrive at a solution which satisfies the largest number of constraints. The above abilities of processing incomplete data and supporting conflict resolution via constraint satisfaction suggest their applicability in scheduling or RS. However, their application to real problem sizes (due to a large number of variables involved in generating a feasible schedule) and real-time/reactive environment was likely to be difficult by using relaxation models (i.e., pre-assembled systems relaxing from input to output along a pre-defined energy contour). In these approaches [5], [109], the neural networks are defined by energy functions. Models that use backpropagation networks, applying the gradient-descent technique in a feed-forward network to change a collection of weights so that the cost function can be minimised, are more developed and much faster than relaxation models. They are applicable in a real-time RS environment [143] (for, e.g., selecting the best performing dispatching rules for all scheduling/sequencing decisions in RS systems with respect to given management objectives, if they are combined with quick simulation for evaluation (see it detailed in next section provided with example methods). The ability to recognise patterns in the current factory state may allow a neural network trained through the results of many experiments to suggest an appropriate response to dispatching rule selection in the non-steady-state factory typical in dynamic manufacturing environment [24].

4.2.6 Genetic Algorithm Based Models

Genetic algorithms are adaptive search techniques which imitate natural selection and genetics. They have been utilised since the late seventies as an approach to solving combinatorial optimisation problems. Their mode of operation as described by [105] is to maintain a set of candidate solutions called a 'population' and to mimic nature by coding a series of parameters to be optimised in the form of a 'chromosome' representing a member of the population and evaluated by a 'fitness function'. The population is initiated at random and evolves in cycles called generations. Starting from a population of chromosomes each of which represents the same series of parameters but with different values, a set of evolution rules is applied in order to arrive at a better fitting population, i.e., where the parameters have reached their optimal value. The evolution is basically performed by genetic operators such as crossover, inversion or mutation. It is then necessary to define a cost function for each chromosome, which depends somehow on the values of the parameters encoded in bit strings of varying length. A combinatorial optimisation problem then results in, consisting of finding the values of the parameters that minimise or maximise a pre-defined cost function. Genetic algorithms have proven to be very efficient in parameter optimisation, classification and learning. Scheduling has recently started to emerge as a new field of application. Genetic algorithms use randomisation techniques to find a local optimum without exhaustively searching through the state solution space. Their main advantage lies in their ability to hop randomly from schedule to schedule, allowing them to escape from local optima in which other algorithms might land. Their applied operators may take into account several real-life constraints, e.g. tool changes, material transport, etc. They are easy to implement and fast to run, and therefore can be taken into consideration as a 'reasoning' methodology for RS in live environment. One of the nice features of genetic algorithms according to [103] is their ability to interface with other approaches for solving scheduling problems.

4.2.7 Stochastic models of proactive scheduling

Stochastic and dynamic manufacturing environments are full with uncertainties due to often unexpected events. The manufacturing system itself as a technical and material processing system is not faultless. There is a certain probability of faults that are foreseeable to occur during the schedule execution and may lead to disruption. To detect faults of real-time systems, the system behaviour must be continually observed and its divergence from the normal behaviour (i.e. predicted execution of an advance schedule) classified from the view of the effects on the execution/performance degradation of further processes [77]. It supposes a close coupling the control and fault detection during schedule execution. Approaches to the latter function are either heuristic or systematic/model-driven. Dominant aspect of the heuristic approach is to describe correct system behaviour with a set of conditions on visible apparent progress of schedule execution needed to correct system behaviour (e.g. using frames assigned to conditions). The clue of the systematic or event-driven approach is to provide a complete model of the whole system describing its (correct) behaviour in detail. A preventive fault detection in the advance schedule execution is performed by checking the system behaviour against the model behaviour [65], and by interpreting every discrepancy as a fault in advance [77]. The essential difference between both classes of fault detection is that the heuristic approach can make use of visible 'alarms' only, and the systematic approach uses system states that are derived from the observations. The advantage of the first is that a system description is fast to create because only some conditions must be formulated without generating a complete and therefore expensive description needed for the systematic approach. But in the latter, completeness of the system model result in the advantage of nearly every effect can be fast detected.

Neglecting the details, the schedule execution model is built by variables and predicates. If the system works correctly, these predicates are satisfied simultaneously. Such a structure consisting of a finite set of variables with given domains/conditions constraining the variables is called a constraint net, the conditions are called constraints. An assignment of one value to each variable such that every constraint is satisfied is called a solution, the search for a solution is a constraint satisfaction problem (CS). Temporal constraint nets are handy means to describe/evaluate behaviour. For monitoring an advance schedule execution and detecting foreseeable disturbances real-time in a manufacturing/schedule execution process, *temporal constraint nets* can be used for modelling [38], [2], [114].

With respect to proactive scheduling problems, they can be formulated as problems of temporal constraints satisfaction. For a more comprehensive modelling, fault diagnosis and learning, ANNs are used in advanced manufacturing systems [204]. As for proactive scheduling [174] in a more complex shop floor control system, neural-networks that aid a fault forecasting process and employ symptomatic search by hypothesis and test are being considered for using symptomatic information for pattern matching. From detected performance trends (collected by a spreadsheet package, e.g. MS Excel in the executed manufacturing processes), trends/events that are likely to lead to a failure/fault in the current schedule execution can be interpreted in advance (e.g. breakdown patterns of resources, lateness patterns of jobs) and failures/performance degradation can be prevented.

4.2.8 Distributed Agent Architectures

Over the past ten years significant research efforts [62], [127], [185], [25], [46], [159] have been devoted to the development and use of Distributed Artificial Intelligence (DAI) techniques. In the eighties, the Distributed AI community has mainly focused its attention on problems where agents contend for computational resources [47], and less attention was paid to distributed factory scheduling [176]. The importance of distributed decision making in factory environments arises from the fact that factories are inherently distributed, and from the need of effective responsiveness to change, both enforcing in the nineties to use DAI for real-time/reactive scheduling in the factory floor. It proposes the provision of intelligence via a federation of co-operative intelligent agents [74]. Each agent would provide expertise in a particular area or would have capability to effect a particular function. The community of agents would collectively work toward the solution of a problem providing mutual assistance when needed.

An agent-based distributed system is a collection of agents that can be viewed as an organisation [57]. An organisation is defined as a composition of a structure and a control regime. The set of possible structures range from strict hierarchies to complex heterarchies [11]. Selected a particular organisational metaphor can provide a framework. For instance, the metaphor of a hierarchical organisation suggests a system in which agents with a wider scope of control have access to more global information. Agents with a more global perspective of overall goals guide agents with a narrower perspective. The selection of an organisational metaphor depends on the complexity and uncertainty embedded in the task of reactive schedule maintenance, to be performed in a dynamic and stochastic environment. According to [11], complexity at three levels should be considered such as: complexity of information, complexity of task and complexity of co-ordination, and should be combined with types of uncertainty (at levels of information, algorithm, environment and behaviour). Complexity and uncertainty seem to be opposing forces in deciding how to structure an organisation for information processing [57]. Different examples for organisational metaphor in existing systems are given, and the scheduling problem decomposition and control mechanisms are investigated in [11].

Using *multiagent architectures* for distributed decision making provides several benefits such as modularising problem-solving knowledge, alleviating the complexity of developing knowledge-based systems by distributing knowledge amongst a group of co-operating agents, or facilitating the integration of heterogeneous knowledge/reasoning mechanisms. However significant problems yet remain such as how to model the expertise of the application domain and incorporate such models within the multiagent system architecture. *Multiagent reasoning architectures* can be examined according to two perspectives : *(a) Primary reasoning at the individual agent level* where the agent's internal structure is described by four entities such as communicating/reactive entity responsible for interactions, rational entity with determined control mechanisms, specialist entity as it is strongly dependent on application, intentional entity including knowledge about itself. *(b) Hybrid reasoning at the group of agents level*, where reasoning depends on choices performed within the group of agents according to three parameters: (i) task and skill distribution, (ii) communication protocols (information exchange is made through a common memory of blackboard as implicit communication, or, it is performed by message sending/explicit communication mechanism), (iii) co-operation and organisation modes describe the manner how a group of agents co-operate for problem solving, which depends on numerous factors.

For concentrating further on the conceptual model of RS related distributed problem solving, it is performed by hybrid reasoning at the group of agents level, and deals with the interaction of heterogeneous agents sharing their knowledge and abilities to co-operate to solve a global problem split into many tasks [176]. By distinguishing five main concepts of distributed systems: the *agent*, the *object*, the *task*, the *control* and the *communication*, they are viewed in RS relation as follows.

Agent is defined as an abstract/physical entity able to act on itself and on its environment, having a partial representation of this environment and able to communicate with other agents. Its behaviour is the result of its world perception, its knowledge, and of the interactions with other agents. It aims at performing a list of *tasks*, which are parts of a global problem of RS. To execute these tasks, the agent may use a set of *objects*. *Control* defines the co-operation between agents, the group organisation, and its evolution. The co-operation is defined by a co-operation degree that ranges from fully co-operative to antagonistic agents. The idea of co-operation often refers to the co-operation between humans in an organisation simplified by the shared knowledge. The aim of the definition of an organisation is to structure the group of agents to solve the task allocation and co-ordination problems. *Communication* between agents depends on the selected protocol, that is the set of rules that specifies the way to synthesise messages to make them significant and correct. The common denominator applicable to all DAI systems is the *inter agent communication protocol*. Further discussion addresses the 'Blackboard Model' of communication and the message passing between agents in a group of agents architecture in what follows.

Blackboard models of DAI

When the agent architecture maps a *Blackboard Model* built around a multiagent, multitask, real- time decision-making, reasoning capabilities are divided among several independent agents or *knowledge sources* (KS), which co-operate by sharing results through the use of a common memory structure called the *Blackboard (BB)*. The agent activities are co-ordinated by a control module (Control KS), which selects the most appropriate agent to be executed given the current state of the BB. Important changes of the BB are notified to the Control KS through events. The main characteristics of BB systems : (1) independence of agents, (2) strong centralised control, (3) event-driven behaviour, all make them relatively close to classical real-time architectures [22]. However, the design of true real-time BB systems usually requires the introduction of concurrent agents execution, which raises a number of important consistency problems. Aspects such as data access serialisation and delays experienced by agents waiting for locked resources become fundamental in such systems (as further detailed, e.g., at the REAKT architecture [102]).

Group organisation of DAI

When the agents architecture maps a *group organisation* structure, where agents model organisational decisional entities, the control system elaborated from the decision received/command for the automation. The communication system responsible for the information exchanges in negotiation/iterative decision making through a local area network, an information system supports the information required by the other sub-systems. An interface system ensures the dialogue with the human operator and the sub-systems' interaction. Here the decision system is based on a representation of the different agents and holds the problem solving algorithms. It detects the nature of the messages received by the communication system and chooses the best suited strategy to be adopted by the other subsystems. It also manages the operating queue and participates in co-operation between the different agents to attempt to process the task at hand [62]. The *dominant inter agent communication protocol* here is message passing for performing negotiations between a number of agents/people involved in different complementing aspects/local goals of RS related decision making, to finally reach an agreement/a common goal. The control and co-ordination of the negotiation process often depicts the control and co-ordination of the scheduling process. The extent to which this negotiation process is comprehended, dictates the degree of opportunism [131] that one is able to design into a scheduling/RS technology to do it more flexible alike human experts' problem solving.

Some dominant inter agent communication protocols

CDPS nets

In a distributed factory and RS decision environment there is an important DAI issue: how decisions should be taken by an intelligent sub-system to achieve its goal in the presence of other intelligent sub-systems with their own goals [185]. Co-operative distributed problem solving (CDPS) [45] relates to how the elements of a loosely coupled network can work together to solve problems (such as reactive scheduling) that are beyond their individual capacities. Each problem solving node in the network is capable of sophisticated problem solving and can work independently, but the problems faced by nodes cannot be solved without co-operation. In CDPS network, each node has sufficient knowledge to at least formulate a partial solution without assistance from other nodes. Inter-node co-operation is often the only way to develop an acceptable overall solution. Each node also has significant expertise in control and communication strategies. The problem is decomposed into sub-problems and the nodes work together on the sub-problems, their individual solutions should fit into a global solution. Thus the nodes must co-ordinate their asynchronous and parallel problem solving activity to produce solutions compatible with their interdependent sub-problems. Nodes must rely on sophisticated local reasoning to decide on appropriate actions/interactions.

Contract nets

Contract nets are used for multistage negotiation in multiagent systems of solving distributed constraint satisfaction problems like RS, by, e.g., [33]. It extends the basic contract net protocol/negotiation paradigm [37] to allow iterative negotiation during bidding and awarding. Another extension of contract nets towards the notion of 'Consortium' (a temporary and logical grouping of agents executing the same task, e.g., a RS task as an enterprise activity) is given by [145]. With a holistic view to schedule generation/execution like 'business processes'/tasks within a CIM-engaged enterprise, the latter extension (based on an integrated agent architecture called HOLOS) provides high control flexibility for negotiation, and supports to envisage virtual manufacturing [75].

DAI architectures enable a scheduling system to be capable of dealing with the right information at the time and place needed. Furthermore, the information should be processed and presented as a recommendation based on incremental decisions by the agents in some framework for negotiation (like the contract net protocol of HOLOS being designed for CIM environment of enterprises, by [144]). The latter offers an infrastructure for RS-related problem solving by an agent architecture with an explicit communication mechanism.

Solution approach to RS	Accommodated model type	Learning embedded	Application system	Name
Opportunistic	Constraint Satisfaction Problem (CSP)		(Fox, 1987, 1994)[56][59]	ISIS-2
			(Fargher et al., 1987)[50]	
			(Farhoodi, 1990)[53]	
			(Berry, 1992)[13]	
			(Collinot et al., 1989)[32]	SONIA
	CSP+BB model of DAI		(Smith, 1994)[164]	
			(Beck, 1993)[11]	OPIS
				TOSCA
			(Sarin et al.,1989)[153]	
	CSP+BB+ operator model	Case-based	(Szelke & Márkus G., 1994)[171]	SUPREACT
	CSP+human preference	Case-based	(Sycara & Miyashita, 1994)[175]	CABINS
	CSP+ DAI framework		(Burke & Prosser, 1991)[21]	DAS
			(Hadavi et al., 1994)[68]	ReDS
Micro-Opportunistic	CSP+BB model		(Hynynen, 1989)[80]	BOSS
			(Sadeh, 1994)[150]	Micro-BOSS
			(Berry, 1992)[13]	
ANN based	ANNs+simulation	supervised	(Chryssolouris et al., 1991)[31]	
		supervised	(Liu and Dong, 1996)[108]	
		supervised	(Rabelo L. et al., 1995)[143]	
		supervised	(Hoong et al., 1991)[78]	
	ANNs	supervised	(Kurbel & Ruppel, 1996)[98]	LEISTAND
		supervised	(Garner & Ridley, 1995)[61]	
		supervised	(Willems & Brandts, 1995)[198]	
		supervised	(Takatori & Kazaku, 1996)[180]	
		supervised	(Garetti & Taisch, 1995)[60]	
		supervised	(Khaw et al., 1991)[92]	
	ANNs+Semi Markov model	supervised	(Yih, 1992)[206]	
	Neuro-fuzzy	supervised	(Bugnon et al., 1995)[20]	FUN
		supervised	(Grabot et al., 1994)[66]	SIPAPLUS
		supervised	(Roy and Zhang, 1996)[148]	
Genetic algorithms based	GA+ANNs+sim.		(Rabelo L. et al., 1995)[143]	
			(Syswerda, 1991)[179]	
	GA		(Fang et al., 1993)[51]	
			(Starkwether et al., 1992)[166]	
			(Mulkens, 1994)[124]	
			(Wiendahl & Garlichs, 1994)[194]	
Fuzzy based	Fuzzy sets	Inductive ML	(Kerr and Kibira, 1994)[90]	
			(Dubois et al., 1993)[44]	
			(Yuan and Wu, 1991)[207]	
			(Dorn et al., 1994)[39]	
			(Schmidt, 1994)[155]	
DAI	Agent architecture		(Sycara et al., 1991)[176]	CORTES
			(Rabelo R.J., 1994)[144]	HOLOS
			(Prosser, 1989)[140]	DAS

TABLE 4.1. Solution approaches to reactive/proactive scheduling

4.3 Solution Approaches - Methods, Techniques, Tools

4.3.1 AI-BASED METHODS AND HEURISTIC SEARCH TECHNIQUES OF RS

Various approaches to reactive scheduling are reflected in the AI and the production management research literature and reviewed in, e.g., [6] and lately in [170]. In the latter overview, some general problem features pervading a number of problem definitions/solutions approaches in the field are also referred to. Reactive scheduling is generally conceived as the real-time revision/repair/maintenance of an advance schedule, (under execution, but flawed at execution time) to keep it in line with the live status of shop floor processes/events. Solution to the problem can be achieved by various AI-based approaches combined with heuristic search techniques.

A classification of different solution approaches to RS based on the comprehensive study of literature can reflect the recent tendencies and major likely trends in relevant aspects of real-time control related RS as envisaged today, and can be found in a summarised form in Table 4.3.1.

Opportunistic incremental constraint satisfaction (CS)

The RS solution process is mostly conceptualised as a contingency-driven, repeated, quick decision making for constraint satisfaction (CS). It is necessary to satisfy, or at least, attempt to satisfy various classes of constraints, ranging from the limited availability of resources (e.g., machines, tools, materials etc.), to the higher management objectives (such as completion dates of orders etc.). Due to the RS problem complexity (outlined in detail in section 1.3), the AI-based or hybrid approaches to RS can be taken into consideration for RS/CS problem solving. These solution approaches generally prefer *an incremental revision/repair* of an advance schedule [164] until it becomes free of conflicts/constraint violations, to *keep continuity* of executed shop floor processes as much as possible.

During the execution time of a schedule, *Constraint Satisfaction Techniques* can achieve this *by either of the two ways: (1) a generative approach* to RS by constructive backtracking to incrementally extend a consistent partial assignment of start time values to the operations until a consistent total value assignment can be reached [50], [151], or *(2) an iterative/incremental repair* of a complete but inconsistent assignment until its conflict free state is achieved [117], [175], [174].

Genetic algorithm based schedule optimisation

A genetic algorithm is developed by [124] for a flow-shop scheduling problem with the criterion of minimising the makespan of the jobs. For two or three machine cases, it was found to perform in a satisfactory manner compared to the well known Johnson's algorithm. The number of jobs and machines involved in the experiments was found to have positive influence on the convergence of the genetic algorithm developed. The larger the problem size the easier it could be solved by the algorithm.

A genetic algorithm for scheduling with resource consumption is reported by [166], and another algorithm for schedule optimisation is given by [179]. They both are used for a compromise analysis process given for reactive scheduling in [143] to find the best schedule from a set of solution candidates (by a crossover operation developed in [179], in order to satisfy given multiple objectives.

ANNs-based schedule re-optimisation

Neural networks have shown good promise for solving some classic, textbook job shop scheduling problems. These implementations have been based on relaxation models (i.e., pre-assembled systems which relaxes from input to output along a predefined energy contour). The neural networks are defined by energy functions in these approaches [5], [109] that are not suitable for solving realistic job shop scheduling problems with multiple objectives. It is even difficult to get a good sub-optimal solution when attempting to solve problems in near real-time.

A system with the sole use of neural network methodology for dynamic scheduling is described by [78]. They consider the dynamic scheduling for job shops as one needing an adaptive approach of 'least commitment' on resource allocation, while integrating objectives/constraints (those responding to market demands) from the strategic/plant level of decision making with those that are given by the current shop status (e.g. distribution of work load, status of WIP inventory). They address the problem of using a neural network to learn about and aggregate the role of various considerations. The possibility is explored of training a neural network to recognise the co-operative and competitive roles of the above factors in contributing to the current objective of the scheduling system. The trained network was then used to continuously monitor the shop status and select/assign jobs to resources.

Another approach with the sole use of ANN, a so called 3-D neural network approach by Khaw [92] is applied for reactive scheduling of multi-dimensional resources (such as materials, machines, manning, money, management information) in the shop floor. In this concept, the scheduling decisions made at one part of the shop floor may often have major influences on decisions at another part, hence, creating a massive web of inter-relationships that are difficult to visualise. If unsolved, these problems may lead to late deliveries, increased costs, inefficiently used resources, poor quality, and insufficient flexibility. Khaw [92] propose a 3-D neural network model for resource allocation in dynamic shop floor environments with multiple objectives. It uses a backpropagation learning algorithm to train a set of existing scheduling data, but does not requires the scheduling variables to be explicitly defined for achieving the objective function. Knowledge representation in the 3-D model is a combination of a distributed representation (where each object/concept is represented by a pattern of features) and a 'localised' representation (where the single nodes are associated with concepts, and causal relationship between concepts is represented by the strength/weights of connection between them). The main activation (transfer) function used in each processing element of the neural network is the hyperbolic tangent function. The developers used 240 training sets and 60 testing sets, where the training sets were obtained from manual scheduling data. Although encouraging results are reported, further research is needed before the real-life applicability of the 3-D model [92].

4.3.2 Combined Methods of Reactive Scheduling

With generally using a simulation framework for schedule analysis/ evaluation, different combinations of AI-based methods, learning techniques and fuzzy tools have been reported in the research and application literature of RS problem solving. They are discussed in next sections.

Methods using ANNs, simulation and fuzzy sets theory

In a *combined approach* to scheduling and reactive scheduling, some tests have been made in [66] to complement it with methodology for making compromises between dispatching rules (used in the simulation-based industrial scheduler) in order to satisfy multiple objectives. The aim of this approach was to build some aggregate rules which provide intermediate results between the results of the elementary rules of which they are composed. Utilisation of different priority dispatching rules strongly depends on the set of jobs which should be analysed. Compromises between rules (surveyed by e.g. [123]) are defined by using fuzzy logic, and each rule weighted in accordance with the characteristics of the jobs considered, in this approach. In order to balance the influence of each elementary rule in relation to manufacturing context, a neural network method was applied for selecting/tuning the weighting values of the rules. Results were compared with the results of another method making fuzzy compromises between rules and using some domain knowledge (not directly possible to introduce into the neural network). Tests have pointed out that a combined use of fuzzy logic and neural networks is likely the best solution to introduce a partial knowledge about the system and, thus, also shorten the learning phase of the ANN.

Another example for such a combined approach is the system called FUN (FUzzy-Neuro) that solves dynamic shop scheduling problems by a model using Neural Networks and fuzzy logic. It is inspired by neuro-fuzzy rules on an allocation system used in computer science ([167] to balance the workload and find solution real-time in function of constraints), and based on minor modifications that adapt the Stoffel's method to the particular scheduling problems of a shop. The problem solving model in FUN is based on a method composed of a static part given by fuzzy logic [208], and a dynamic part that comes from a neural network. The dynamic adaptation of the fuzzy controller by a back propagation network allows real-time reaction in case of damages or perturbations.

ANNs, genetic algorithms and simulation

A knowledge-based reactive scheduling methodology with integrated use of Neural Networks, Simulation, Genetic Algorithms and Machine Learning (induction) in RS problem solving is implemented in a prototype system for job-shop/flow-shop scheduling/reactive scheduling to real-time industrial environments by [143]. Their methodology can exploit the parallel processing and modelling capabilities of the neural nets, simulation and genetic algorithms, and can learn scheduling rules from the resulting decisions that may satisfy multiple objectives. It presumes availability of a large list of known dispatching rules and generates a small set of candidate rules that best perform with respect to at least one of major performance measure such as job tardiness, makespan, etc. This analysis is carried out by ANNs, developed by Rabelo (1995) and Chryssolouris (1991) [143], [31], trained to rank the available rules, and extended to be usable in real-time regime. Each of the candidate rules is then concurrently evaluated by a statistical analysis via real-time simulation-runs to determine its impact on the future evolution of executed processes from the current state of the system. The above simulation engine is also utilised by the Genetic Algorithm Process (GAP), which generates a new schedule from the simulated candidate schedules to combine the best features of the most attractive rules and simultaneously achieve good performance for all objectives. This GAP uses an adaptive mutation [143], exchanging two arbitrary jobs' position, and exhibiting high flexibility to develop the schedule (with regard to fixed and in-process operations/jobs already in the manufacturing process). In addition, for the knowledge extraction contained in the schedule resulting from the GAP, an Inductive Learning Algorithm, called Trace-Driven Knowledge Acquisition (TDKA, developed by [205]), is used for deriving a "new rule". This rule, although being not a simple dispatching rule, can be used to re-generate schedules in the same way that other dispatching rules like SPT are used.

Case based reasoning/learning with rule based reasoning and simulation

Case-based reasoning/learning (CBR/L) [95] is a general paradigm for reasoning from prior learned experience/empirical knowledge. It assumes a memory model for representing, indexing and storing past cases of a problem solution in an organised form in the case-base (CB), and a process model for retrieving and modifying old cases to be applicable for a new similar problem instance, while assimilating new cases in the CB. AI researchers seek to understand the nature of human thought, and examine a range of human cognitive behaviour including memory, learning planning and problem solving. In these, CBR/L provides a cognitive model [95]. Learning, missed by rule based systems, is an inherent feature of CBR. CBR/L provides a basis for intelligent systems to solve problems by re-use/adaptation of past solutions to new problem situations.

An example system for using CBR/L for schedule revision and reactive repair based optimisation in manufacturing is called CABINS (reported by [175] and [119] that interactively acquires user's scheduling preferences in the form of cases and re-uses them to reactively manage complete but flawed schedules in response to unexpected environmental events. In the knowledge acquisition phase, the user interacts with the CABINS to select repair heuristics for adapting the search procedure to the search space features, evaluate the acceptability of repair by the human, and explain unacceptable repair results. In the performance phase, the results are used in a constraint based schedule repair process, which integrates CBR with rule-based constraint propagation techniques to do time-dependent and resource constrained reasoning. Case-indexing in CABIN applies some global/local context features relevant to global optimality/local feasibility of a repair in a given situation. Retrieval by a serial search in flat memory organisation is performed with a k-nearest-neighbour matching process. Test experiments with the system with respect to various optimum criteria has shown that the CB approach can reliably acquire user preferences without a causal model and outperform other repair methods in schedule revision.

Another example for CBR/L-based reactive scheduler is a system that is developed with a supervisory perspective of control and called SUPREACT [172], [171], [173]. For integrating human with computational agents (KSs) in real-time control, it uses a cognitive operator function model and a corresponding blackboard architecture. CBR/L, rule-based and model-based agents interact via the blackboard to resolve RS related conflict situations, by using human preferred choices from solution alternatives during the interactive schedule repair. The hybrid approach (combined with simulation) is also considered for proactive schedule maintenance [174].

Inductive learning in combined use with simulation and fuzzy sets

A SIMAN application with combined use of an inductive learning tool C4.5 for interactive reactive scheduling is reported by Kerr and Kibira (1994) [90]. They aimed at carrying out some experiments to demonstrate the feasibility of applying machine induction for learning reactive scheduling strategies. With this experimental system, RS knowledge is acquisited in live manufacturing practice from a skilled human expert of scheduling a telephone production line. The simulation model of the plant is used to log the scheduling decisions of the experienced human scheduler, and machine learned decision rules can then achieve the schedule/control of the production line by entirely taking over the task of the human scheduler. Learned decision rules enable to meet some goals like target buffer levels and volume throughput by the end of every shift. In the combined system for learning and using learnt RS strategies by an inductive learning tool (C4.5 [142], a descendant of ID3) is embedded together with fuzzy tools for RS into the simulation framework used for experiments to gain

RS strategies adequate to the control environment (a telephone production line, a flow-shop).

Fuzzy sets combined with other techniques

Work is reported in [207] on a fuzzy dynamic programming approach, combining methods of fuzzy set theory, fuzzy reasoning and mathematical programming for handling uncertainties in real-time scheduling of automated guided vehicles (AGVs) in FMS. The algorithm used in this approach provides a set of statistical parameters obtained from the off-line testing of each transportation path in an AGV system. The complexity of the AGV scheduling problem lies in the fact that travelling time is not just a function of geometric distance of load/unload/machining centres, but heavily influenced by layout factors and actual occupation conditions in the shop floor. All of these physical factors may be treated as model uncertainties, and a fuzzy set of transportation times is used in the approach to describe fluctuations. The algorithm for path programming is derived subject to these fuzzy relations and constraints, and is handled by fuzzy dynamic programming. The advantages of this approach come from the fact that the domain of the fuzzy set is continuous, allowing easy adjustment of the model by tuning parameters. The model parameters are also adjustable to working conditions and to any change in the system. Collision, a major source of conflict in AGV scheduling is manageable by this approach, also in multi-AGV systems.

Kerr and Walker [91], Dorn [39], and also Schmidt [155] draw attention to the fact that predictive schedules themselves can be made more robust against contingencies if the imprecision inherent in the scheduling parameters is explicitly represented. They propose fuzzy arithmetic as a means of achieving this by propagation of fuzzy temporal constraints through the schedule. These approaches allow for the propagation of uncertainty in such parameters as processing times, material arrival times etc., to dependent events so that it accumulates in such a way that reflects how knowledge of event timings becomes increasingly imprecise as the number of dependencies increases. The advantage of fuzzy temporal constraint propagation lies in its potential for decoupling different regions of the schedule which, because of imprecision in the parameters, may be regarded as non interacting. This is an important issue in RS, in which the effects of unexpected events/contingencies can be localised to a region in which they have a dominant impact, rather than being propagated out of the limits of the known horizon where in reality, they would be swamped by uncertainty arriving from other sources. A complementary application lies in the use of this approach to protect schedules against the effects of uncertainty as has been investigated by Chiang and Fox [29] for the case of machine breakdown.

4.4 Conclusions: Future Research Issues in the Field

The bulk of information technology tools for factory scheduling currently in use in factories have deficiencies regarding adequate representation, scheduling power, human-computer interaction, reactivity, flexibility and ease in modelling constraints and objectives. Research in knowledge-based (reactive) scheduling, surveyed in [170], has addressed these issues over the past few years, and results have started to disseminate into industry [10] [88].

The majority of approaches to the problem and application systems in the field of RS are currently in the experimental stage, except some matured systems like OPIS, ReDS, DAS, Micro-BOSS, etc. They tend to make much more emphasis on the revised/repaired schedule quality than the response time efficiency of RS problem solving. Efficiency concerns are addressed in most solution approaches in an extent that disruption to the execution of the advance/original schedule is minimised by incrementally revising the invalidated part of the schedule. However, the more pragmatic problem of real-time reacting to a change within a specified time frame have not been fully addressed, except some special real-time framework for RS like REAKT [102]. The pragmatics of timely reaction is a fundamental point in real-time interactive systems, and more recently in distributed systems with co-operating intelligent agents. These systems demand a framework of the type known from general reactive planning architectures [54], for prioritising and managing pending conflicts [74].

Ongoing research, as projected by [88] and [68], has to address such issues as: (1) how to ensure generality of knowledge representation to broader applicability of an RS system; (2) which representation amongst recent proposals for a universal, generic factory model are most effective for describing possible behaviours of the factory floor environment, and for the implementation of efficient constraint propagation schemes; (3) how to integrate predictive and proactive/reactive scheduling decision making in the same knowledge based system; (4) how to asses/learn thresholds basing for decisions on problems when to be predictive (i.e. re-schedule from scratch), when to be reactive (revise/repair only the flawed parts of a schedule), and when to initiate proactive revision of a current schedule. The provision of answers to such questions requires considerable amount of further research efforts in the exploration of alternative schemes for the representation of domain knowledge of reactive operation management in manufacturing processes and the best ways of building it into the decision making processes.

Further research has been envisaged in creating correct stochastic preventive action models [41] or correct behavioural models [77] to represent the evolving behaviour of the controlled (manufacturing) system for diagnosing the failure states, with discriminating whether the complete re-scheduling from scratch/RS/PS would be the adequate control response in the mirror of the current system state.

Some researchers in the field (such as, e.g., [143], [84]) envisaged the possibility of blending different approaches to derive better systems for RS and real-time production control in terms of all aspects such as capability, applicability, versatility, expandability and speed. Hybrid systems are receiving greater attention from researchers because they work to overcome each other's limitations while gaining strength from each other. This does not mean that they are free from problems. The main problem that one encounters while integrating these different approaches is that of increased complexity. In an attempt to model the controlled system complexity with a degree of fidelity to live operating manufacturing systems, the models themselves become highly complex, in turn opening a plethora of related IT issues.

Another perspective of future investigations has to address the proper macro environment and IT demands for distributed scheduling in a real-time regime. Namely, it has to deal with further investigation of agent architectures and also the co-operation and control in them to meet the new requirements for flexibility and responsiveness of scheduling (by, e.g., permitting the addition/removal of site agents). These requirements are being imposed upon manufacturing scheduling systems by the rapid spread of lean, agile and holonic manufacturing across the manufacturing industry [145].

A comprehensive model structure which can adequately reflect all the principal functions of a manufacturing system, and work as an effective framework [165] may be the solid basis on which an effective real-time control and RS system can be constructed. Techniques such as simulation, Petri nets, fuzzy logic, neural networks and hybrid knowledge based systems rely on these model frameworks. Limitations exist in the above techniques themselves. Simulation is a trial-and-error tool unless combined with some form of intelligence, and even that, computation time remains a hurdle in real-time environment. Petri nets have proved effective for performance evaluation, yet their effectiveness for real-time scheduling and control has not been adequately explored. Fuzzy logic has not been explored to its fullest potential to be applied for RS. Neural networks cannot provide optimal decisions, but their inherent learning ability makes them ideally suited for rapidly changing systems. Integrating neural nets with knowledge based systems and simulation techniques seems to have shown a lot of promise [160].

With increasing trends for AI tools and generic modelling frameworks to permeate live manufacturing environments, methods to elicit, represent, store/retrieve and process the control/ scheduling domain knowledge and tools to configure situation adequate control/RS strategies, have matured to the point of starting to reduce the gap between the theoretical results of knowledge based reactive scheduling and their broader industrial application. The issue of real-time control including RS has remained important for both academic research and industrial application interest, as it is indicated by the abundant literature documenting both purely academic endeavours and industry-sponsored research efforts. While the ultimate objective may be to develop a 'plug and play' type of RS/PS supported real-time production control system with a high degree of implementability, reliability and ease of use in diverse industrial settings, this cannot be achieved unless the large number of theoretical limitations in currently existing approaches are overcome, or more capable methods developed.

4.5 References

[1] Ahuja, S.J., and Valavanis, K.P., 1988, Modified Petri nets for comprehensive modelling of flexible manufacturing systems. In: *Proc. for IMACS'88*, Vol. 4, pp. 532-534.

[2] Alur, R., Courcoubetis, C., and Dill, D., 1991, Model-checking for probabilistic real-time systems. *Proc. of the 18th Int. Colloq. on Automata, Languages and Programming*, Madrid, July 1991, Springer Verlag, Lecture Notes in Computer Science 510, pp.115-126.

[3] Andersin, H.E., 1992, Performance measurement as an integrating link between man and CIM. In: *Human Aspects in Computer Integrated Manufacturing*, Eds.: G.J. Olling and F. Kimura, Elsevier Sci. Publishers, IFIP, 1992, Tokyo, pp. 495-513.

[4] Arai, E., Amnuay, S.T. and Uchiyama, N., Real-time simulation and monitor forecast for dynamic scheduling in distributed production systems. *Proceedings of JSPE-IFIP WG 5.3 Workshop DIISM'93*, Japan, 1993, pp. 137-147.

[5] Arizono, I., Yamamoto, A., and Ohta, H., 1992, Scheduling for Minimising Total Actual Flow Time by Neural Networks. *Int. J. of Production Research*, Vol. 30, No.3, pp.503-511.

[6] Atabakhsh, H., 1991, A survey of constraint based scheduling systems using an AI approach. *Artificial Intelligence in Engineering*, Vol. 6, pp. 58-73.

[7] Avouris, N.M. and Gasser, L. (Eds.), 1992, *Distributed Artificial Intelligence: theory and practice*, Kluwer Academic Publisher.

[8] Baker, K. R., 1974, *Introduction to Sequencing and Scheduling*, New York: John Wiley & Sons, 1974.

[9] Barcio, B.T., Ramaswamy, S., and Barber, K.S., 1996, An Object-Oriented Modeling and Simulation Environment for Reactive Systems Development. *J. of Flexible Manufacturing Systems*, Kluwer Academic Publishers, Norwell, USA.

[10] Basaglia, G., and Guida, M., 1991, Operational Research and Artificial Intelligence in the Development of Real-time Scheduling Systems, the BIS Esprit Project. *IJCA We'91: Workshop on Advances in interfacing production systems with the real world.* Sydney, Australia, 1991.

[11] Beck, H.A., 1993, TOSCA: A novel approach to the management of job-shop scheduling constraints. In: *Realising CIM's Industrial Potential*, Eds.: C. Kooij, P.A. Mac Conail and J. Bastos, IOS Press, 1993, pp.138-149.

[12] Ben-Arieh, D., 1992, Concurrent modeling language (CML) for discrete process modeling, simulation and control. *Journal of Intelligent Manufacturing*, (1992), 3, pp.31-41.

[13] Berry, P.M., 1992, Scheduling: A problem of decision making under uncertainty. *Proceedings 10th European Conference on Artificial Intelligence, ECAI'92*, Vienna, John Wiley & Son pp.638-642.

[14] Bohner, P., 1995, A multi-agent approach with distributed fuzzy logic. *Computers in Industry*, Vol. 26, pp. 219-227.

[15] Brandimarte, P., 1992, Neighbourhood search-based optimization algorithms for production scheduling: a survey. *Computer Integrated Manufacturing Systems*, Vol. 5, No.2, pp.167-176.

[16] Brandimarte, P. and Villa, A. (Eds.), 1995, *Optimization Models and Concepts in Production Management*, Gordon and Breach Publishers SA, Basel, Switzerland.

[17] Brown, M.C., 1988, The dynamic re-scheduler: conquering the changing production environment. *Proceedings 4th IEEE Conference on AI Applications*, San Diego, USA, pp.175-180.

[18] Bruno, G., and Biglia, P., 1985, Performance evaluation and validation of tool handling in FMS using Petri nets. *Proceedings of Int. Workshop on Timed Petri Nets*, Torino, Italy, July 1985, pp.64-71.

[19] Bruno, G., and Marchetto, G., 1986, Process-translatable Petri nets for the rapid prototyping of process control systems. *IEEE Transactions on Software Engineering*, Vol. SE-12, February 1986, pp.346-357.

[20] Bugnon, B., Stoffel, K. and Widmer, M., 1995, FUN: A dynamic method for scheduling problems. *European Journal of Operational Research*, Vol. 83, (1995), pp. 271-282.

[21] Burke, P., and Prosser, P., 1991, A Distributed Asynchronous System for Predictive and Reactive Scheduling. *Artificial Intelligence in Engineering*, Vol.6, No.3, pp.106-124.

[22] Carver, N., and Lesser, V., 1994, Evolution of Blackboard Control Architectures (Invited paper), *Expert Systems With Applications*, Vol. 7, pp. 1-30.

[23] Caselli, S., Papaconstantinou, C., Doty, K.L., and Navathe, S., 1992, A structure-function-control paradigm for knowledge-based modeling and design of manufacturing workcells. *Journal of Intelligent Manufacturing*, (1992), 3, pp.11-30.

[24] Caskey, K. and Storch, R.L., 1996, Heterogeneous dispatching rules in job and flow shops. *Int. J. Production Planning and Control*, Vol. 7, No. 4, pp. 351-361.

[25] Chaib-Draa, B., Moulin, B., Mandiau, R., and Millot, P., 1992, Trends in Distributed Artificial Intelligence. *Artificial Intelligence Review*, 1992, No. 6, pp.35-66.

[26] Chang, F.C., 1985, A knowledge-based real-time decision support system for job-shop scheduling at the shop-floor level. *PhD Dissertation*, Ohio State University, USA.

[27] Chase, Ch. and Ramadge, P. J., 1992, On Real-Time Scheduling Policies for Flexible Manufacturing Systems, *IEEE Transactions on Automatic Control*, Vol. 37, No. 4, April 1992, pp. 491-496.

[28] Chengen, W., Jianying, Z., and Zhongxin, W., 1993, An expert system for FMS control, *Intelligent Systems Engineering*, Winter 1993, pp. 223-230.

[29] Chiang, W.-Y., and Fox, M.S., 1990, Protection against uncertainty in a deterministic schedule. In: *Proc. of the Fourth Int. Conference on Expert Systems in Production and Operations Management*, May, 1990, Hilton Head Island, S.C., USA.

[30] Chiuc, C., and Yih, Y., 1995, A learning based methodology for dynamic scheduling in distributed manufacturing systems. *Int. J. of Production Res.*, Vol.33, No.11, pp. 3217-3232.

[31] Chryssolouris, G., Guillot, M., and Domroese, M., 1991, The use of neural networks in determining operational policies for manufacturing systems. *J. of Manufacturing Systems*, Vol. 10, No. 2, pp. 166-175.

[32] Collinot, A., and Le Pape, C., 1989, Testing and Comparing reactive scheduling strategies. *Proc. 1989 AAAI-SIGMAN Workshop on Manufacturing Production Scheduling*, AAAI-SIGMAN, Detroit, USA.

[33] Conry, S., Meyer, R., and Lesser, V., 1988, Multistage negotiation in distributed planning, *Readings in Distributed Artificial Intelligence*, Morgan-Kaufmann, 1988, pp. 367-384.

[34] Costello, D., Jordan, P., and Browne, J., A knowledge-based tool for reactive scheduling. In: *Artificial Intelligence in Reactive Scheduling*, E. Szelke and R. Kerr (Eds.), Chapman & Hall, London, 1995, pp.95-114.

[35] Cras, J-Y., 1993, A Review of Industrial Constraint Solving Tools - AI Perspectives. *Report published by Artificial Intelligence*, Oxford, UK, 1993, ISBN 1 898804 001.

[36] Croce, F.D., 1995, Generalized pairwise interchanges and machine scheduling. *European Journal of Operational Research*, Vol. 83, pp. 310-319.

[37] Davis, Randall, Smith, Reid, 1983, Negotiation as a metaphor for distributed problem solving, *Artificial Intelligence*, No. 20, 1983, pp. 63-109.

[38] Dechter, R., Meiri, I., and Pearl, J., 1991, Temporal constraint networks. *Artificial Intelligence*, Vol.49, pp.61-95.

[39] Dorn, J., Kerr, R.M., and Thalhammer, G., 1994, Reactive Scheduling in a Fuzzy Temporal Framework. In: E. Szelke, and R.M. Kerr, (Eds.), *Knowledge Based Reactive Scheduling*, IFIP, North-Holland, Amsterdam, 1994, pp.39- 56.

[40] Drummond, M., and Bresina, J., 1990, Anytime Synthetic Projection: Maximizing the Probability of Goal Satisfaction. *Proceedings of Conf. AAAI-90*, AAAI Press, Menlo Park, Calif., pp.138-144.

[41] Drummond, M., Swanson, K., and Bresina, J., 1994, Robust Scheduling and Execution for Automatic Telescopes. In: M. Zweben and M.S. Fox, *Intelligent Scheduling*, Morgan Kaufmann, San Francisco, California, 1994, pp.341-369.

[42] D'Souza, K., and Khator, S.K., 1994, A survey of Petri net applications in modeling controls for automated manufactruring systems. Survey. *Computers in Industry*, 24, (1994), pp.5-16.

[43] Dubois, D. and Stecke, K. E., 1983, Using Petri nets to represent production processes, *22nd IEEE Conference on Decisions and Control*, pp.1062-1067.

[44] Dubois, D., Fargier, H., and Prade, H., 1993, Handling flexibility and uncertainty in job-shop scheduling. *Proc. FLAI'93 European Workshop on Fuzzy Logic in Artificial Intelligence* (Linz, Austria) pp. 13-17.

[45] Durfee, E.H., 1989, Trends in co-operative distributed problem solving. *IEEE Trans. on Knowledge and Data Engineering*, Vol. 1, No. 1, March 1989, pp. 63-83.

[46] Durfee, E.H. (Ed.), 1991, Distributed Artificial Intelligence, *IEEE Trans. on Systems, Man and Cybernetics*, Vol. 21, No. 16, November 1991, pp. 1167-1183.

[47] Durfee, E.H., Lesser, V., and Corcill, D.D., 1987, Coherent Cooperation among Communicating Problem-Solvers. *IEEE Transactions on Computers*, Vol. C-36, No.11, November 1987, pp.1275-1291.

[48] Durfee, E. H., and Lesser, V. R., 1988, Incremental Planning to Control Time-Constrained Blackboard Based Problem Solver, *IEEE Trans. on Aerospace and Electronic Systems.*, Vol. 24, No.5, Sept. 1988, pp. 647-662.

[49] Ehlers, E.M. and van Rensburg, E., 1996, An Object-Oriented Manufacturing Scheduling Approach, *IEEE Transactions on SMC- Part A: Systems and Humans*, Vol.26, No. 1, January, 1996, pp. 17-26.

[50] Elleby, P, Fargher, H. E., and Addis, T.R., 1988, Reactive constraint-based job-shop scheduling. In: Oliff, M.D. (Ed.), *Expert Systems and Intelligent Manufacturing*, North-Holland, New York, pp. 1-10.

[51] Fang, H-L., Ross, P. and Corne, D., 1993, A Promising Genetic Algorithm Approach to Job-Shop Scheduling, Rescheduling, and Open-Shop Scheduling Problems. *Proc. of the Fifth Int. Conf. on Genetic Algorithms*, pp. 375-382.

[52] Fanti, M.P., Figalli, G., Maione, B., Piscitelli, G., and Turchiano, B., 1990, Discrete-event modelling of Flexible Manufacturing Systems, *Proc. for COMCONEL 90: Communication, Control, and Electronics Conference*, Cairo, Egypt, pp. 46-50.

[53] Farhoodi, F., 1990, A knowledge based approach to dynamic job-shop scheduling. *Int. J. of Computer Integrated Manufacturing*, Vol.3, 1990, pp. 84-95.

[54] Firby, R.J., 1987, An investigation into reactive planning in complex domains. *Proc. Sixth National Conference on Artificial Intelligence*, AAAI-87, American Association for AI, Seattle, USA, 1987, pp.202-206.

[55] Foo, S.Y., Takefuji, Y., and Szu, H., 1994, Job-shop scheduling based on modified Tank-Hopfield linear programming networks. *Engineering Applications of Artificial Intelligence*, Vol. 7, No. 3, pp. 321-327.

[56] Fox, M.S., 1987, Constraint-Directed Search: A case study of Job-Shop Scheduling. *Intelligent Scheduling*, M.Zweben and M.S. Fox (Eds.), Morgan Kaufmann, San Francisco, California.

[57] Fox, M.S., 1988, An Organizational View of Distributed Systems. In: *Readings in Artificial Intelligence*, Eds.: A.H. Bond and L. Gasser, Morgan Kaufmann, San Francisco, 1988.

[58] Fox, M.S., Sadeh, N., and Baykan, C., 1990, Constrained Heuristic Search. *Proc. of the Workshop on Innovative Approaches to Planning, Scheduling and Control*, San Diego, California, DARPA Publ., pp. 309-315.

[59] Fox, M.S., 1994, ISIS: A Retrospective. In: *Intelligent Scheduling*, M.Zweben and M.S. Fox (Eds.), Morgan Kaufmann, San Francisco, California, pp. 3-28.

[60] Garetti, M., and Taisch, M., 1995, Using neural networks for reactive scheduling. In: *Artificial Intelligence in Reactive Scheduling*, R. Kerr and E. Szelke (Eds.), Chapman & Hall (IFIP), London, pp. 146-155.

[61] Garner, B.J., and Ridley, G.J., 1995, Knowledge acquisition for reactive scheduling. In: *Artificial Intelligence in Reactive Scheduling*, Eds.: R.M. Kerr and E. Szelke, Chapman & Hall, London, 1995, pp.156-164.

[62] Gasser, L., and Huhns, M.N. (Eds.), 1989, *Distributed Artificial Intelligence*, Vol. II., Morgan Kaufmann Publishers, San Mateo, California.

[63] Gershwin, S.B., 1989, Hierarchical Flow Control: A Framework for Scheduling and Planning Discrete Events in Manufacturing Systems. Invited paper. *Proc. of the IEEE*, Vol. 77, No.1, January, 1989, pp.195-209.

[64] Goldberg, D.-E., 1989, Genetic Algorithm in search, optimisation and machine learning, *Reading*, 1989.

[65] Goyal, S.K., Metha, K., Kodali, R., and Deshmukh, S.G., 1995, Simulation for Analysis of Scheduling Rules for a Flexible Manufacturing System. *Integrated Manufacturing Systems*, Vol.6, No. 5, pp. 21-26.

[66] Grabot, B., Geneste, L. and Dupeux, A., 1994, Multi-heuristic scheduling in SIPAPLUS: three approaches to tune compromises. *Journal of Intelligent Manufacturing*, Vol.5, pp. 303-313.

[67] Gregor, M., and Kosturiak, J., 1996, ARENA: Environment for Simulation Model Building, *Technical Report of DIE*, University of Zilina, Slovakia.

[68] Hadavi, K.C., 1994, ReDS: A Real Time Production Scheduling System from Conception to Practice. In: *Intelligent Scheduling*, Eds.: M. Zweben and M.S. Fox, Morgan Kaufmann, San Francisco, 1994, pp.581-604.

[69] Harel, D. and Pnueli, A., 1985, On the development of reactive systems. In: *Logistics and Models of Concurrent Systems*, Ed.: Apte, K.R., Springer Verlag, NewYork, pp.477-498.

[70] Harmonosky, C.M., and Robohn, S.R., 1991, Real-time scheduling in computer integrated manufacturing: a review of recent research. *International Journal of Computer Integrated Manufacturing*, Vol. 4, No.6, pp.331-340.

[71] Hasle, G., and Smith, S.F., 1995, Directing an opportunistic scheduler: an empirical investigation on reactive scenarios. In: *Artificial Intelligence in Reactive Scheduling*, R. Kerr and E. Szelke (Eds.), Chapman & Hall (IFIP), London, 1995, pp. 1-11.

[72] Hatono, I., Yamagata, K., and Tamura, H., 1991, Modeling, and On-Line Scheduling of Flexible Manufacturing Systems Using Stochastic Petri Nets. *IEEE Transactions on Software Engineering*, Vol. 17, No. 2, February 1991, pp.126-132.

[73] Hayes-Roth, B., 1985, A blackboard architecture for control. *Artificial Intelligence*, Vol.26, 1985, pp.251-321.

[74] Hayes-Roth, B.,1990, Architectural foundations for real-time performance in intelligent agents, *Journal of Real-Time Systems*, Vol.2, 1990, pp.99-125.

[75] Hitchcock, M., 1994, Virtual Manufacturing - A Methodology for Manufacturing in a Computer. *Proc. of Workshop on the Automated Factory of the Future: Where do we go from here? / IEEE 1994 Int. Conf. on Robotics and Automation*, San Diego, 1994.

[76] Ho, Y.C., 1987, Performance Evaluation and Perturbation Analysis of Discrete Event Dynamic Systems. *IEEE Trans. on Automatic Control*, Vol. AC-32, no.7, pp. 563-572.

[77] Holloway, L.E., Paul, C.J., Strosnider, J.K., and Krough, B.H., 1991, Integration of behavioral fault detection models and an intelligent reactive scheduler. *Proc. of IEEE Int. Symposium on Intelligent Control*, August 1991, Arlington, Virginia, USA, pp. 134-139.

[78] Hoong, L.W., Yeo, K.T., and Sim, S.K., 1991, Application of neural networks in a job-shop environment. *Proceedings of Int. Conference on Computer Integrated Manufacturing, ICCIM'91, 'Manufacturing enterprises of the 21st Century'* , (World Scientific Publ., Singapore), pp. 559-562.

[79] Huhns, M.N., *Distributed Artificial Intelligence*, Research Notes in AI, London, Pitman, 1987.

[80] Hynynen, J.E., 1989, BOSS - An Artificial Intelligent system for distributed factory scheduling. In: Kimura, F. and Rolstadas, A. (Eds.), *Proc. of CAPE'89 IFIP Conference on Computer Applications in Production and Engineering*, Elsevier Science Publ., Tokyo, pp.667-677.

[81] Ippolito, R., Rosetto, S., Vallauri, M., Villa, A., 1986, The emergence of Artificial Intelligence appliactions in manufacturing system control. *IEEE-H2282-2/86*, pp. 472-476.

[82] Jain, S., Barber, K., and Osterfeld, 1990, Expert simulation for on-line scheduling, *Communication ACM*, Vol.33, No.10, Oct. 1990, pp.54-59.

[83] Jensen, K., 1992, Coloured Petri Nets: Basic Concepts, Analysis Methods and Practical Use *(EATCS Monographs on Theoretical Computer Science)*, Springer-Verlag, New York, 1992.

[84] Jones, A., Rabelo, L., and Yih, Y., 1995, A hybrid approach to real-time sequencing and scheduling. *Int. J. of Computer Integrated Manufacturing*, Vol. 8, No. 2, pp. 145-154.

[85] Kaebling, L.P., 1987, An architecture for intelligent reactive systems. In: *Reasoning About Actions and Plans*, Eds.: Michael, P.G. and Lansky, A.L., Morgan Kaufmann, 1987, pp. 395-410.

[86] Karsiti, M.N., Cruz Jr., B., and Mulligan Jr., J.H., 1991, Performance Forecasts as Feedback for Schedule Generation. *Journal of Manufacturing Systems*, Vol.11, No. 5, pp.326-333.

[87] Kasturia, E., DiCesare, F., and Desrochers, A.A., 1988, Real-time control of multilevel manufacturing systems using coloured Petri nets. In: *Proc. of the IEEE Int. Conf. on Robotics and Automation*, Philadelphia, pp. 1114-1119.

[88] Kempf, K., Russel, B., Sidhu, S., and Barett, S., 1991, AI-based schedulers in manufacturing practice. *AI Magazine*, Vol. 11, pp 46-56.

[89] Kerr, R.M., 1991, *Knowledge-Based Manufacturing Management*, Addison-Wesley, Sydney.

[90] Kerr, R.M., and Kibira, D., 1994, Simulation and Machine Induction with Fuzzy set theory, Wien, June 1994, *Proc. of the 2nd IFAC/IFIP/IFOR Conf. on 'Intelligent Manufacturing Systems'- IMS'94*, June 1994, Vienna, Austria, Techn. Univ. of Vienna, pp.397-404.

[91] Kerr, R.M., and Walker, R.N., 1989, A job shop scheduling system based on fuzzy arithmetic. *Proc. of 3rd Int. Conf. on Expert Systems and the Leading Edge in Production and Operations Management*, Hilton Head Island, pp. 433-450.

[92] Khaw, J., Siong, L.B., Lim, L., Yong, U., Jui, S.K., and Fang, L.C., 1991, Shop floor scheduling using a three dimensional neural network model. *Proc. of the Int. Conference on Computer Integrated Manufacturing, ICCIM'91 on 'Manufacturing Enterprises of the 21st Century'*, World Scientific Publ., Singapore, pp. 563-566.

[93] Kirn, S. and Schneider, 1994, STRICT: A Blackboard-Based Tool Supporting the Design of Distributed PPC Applications. *Expert Systems With Applications*, Vol. 7, pp.131-146.

[94] Kohonen, T., 1988, An introduction to neural computing. *Neural Networks*, No.1, pp. 3-16.

[95] Kolonder, J. (Ed.), 1993, *Case-Based Reasoning*, Morgan Kaufmann, San Francisco.

[96] Korf, R.E., 1988, Real Time Heuristic Search, Proc. of AAAI 88, *The Seventh National Conference on Artificial Intelligence*, Aug. 21-26, 1988, Saint Paul, Minnesota, USA, Vol. 1, pp.139-144.

[97] Krogh, B.H., and Sreenivas, R.J., 1987, Essentially decision free Petri nets for real-time resource allocation. In: *Proc. of the IEEE Int. Conf. on Robotics and Automation*, Raleigh, NC, pp.1005-1011.

[98] Kurbel, K., and Ruppel, A., 1996, Integrating intelligent job-scheduling into a real-world production scheduling system, *J. of Intelligent Manufacturing*, Vol. 7, 1996, pp. 373-377.

[99] Kusiak, A., and Villa, A., 1987, Architectures of expert systems for scheduling flexible manufacturing systems. *Proc. IEEE Int. Conf. on Robotics and Automation*, Releigh, USA, 1987, pp.113-117.

[100] Kwok, A., and Norrie, D., 1993, Intelligent agent systems for manufacturing applications. *J. of Intelligent Manufacturing*, Vol.4, pp.285-293.

[101] Laffey, T.J., Cox, P.A., Schmidt, J.L., Kao, S.M., and Read, J.Y., 1988, Real-time knowledge-based systems. *AI Magazin*, Vol. 9, 1988, pp.27-45.

[102] Lalanda, P., Charpillet, F., and Haton, J-P., 1992, A real-time blackboard-based architecture, *Proceedings 10th European Conf. on Artificial Intelligence - ECAI'92*, Vienna, Austria, August 1992, Ed.: B. Neumann, Wiley & Sons, Vienna, 1992, pp. 262-266.

[103] Lawton, G., 1992, Genetic algorithms for schedule optimisation. AI Expert, Vol. 5, pp.23-27.

[104] Le Pape, C., 1994, Scheduling as Intelligent Control of Decision-Making and Constraint Propagation. In: *Intelligent Scheduling*, Eds.: M. Zweben and M.S. Fox, Morgan Kaufmann, San Francisco, 1994, pp.67-98.

[105] Liepins, G.E., and Hilliard, M.R., 1989, Genetic algorithms: foundations and applications. *Annals of Operations Research*, Vol. 21, pp. 31-43.

[106] Likkas, G.P., Avouris, N.M. and Papakonstantinou, G., 1995, Development of distributed problem solving systems for dynamic environments, *IEEE Transactions SMC*, Vol. 25, No. 3, March 1995, pp. 400-414.

[107] Liu, B., 1993, Knowledge-Based Factory Scheduling: Reasource Allocation and Constraint Satisfaction. *Expert Systems With Applications*, Vol.6, pp. 349-359.

[108] Liu, H., and Dong, J., 1996, Dispatching rule selection using artificial neural networks for dynamic planning and scheduling. *J. of Intelligent Manufacturing*, Vol. 7, pp. 243-250.

[109] Lo, Z. and Bavarian, B., 1991, Scheduling with Neural Networks for Flexible Manufacturing Systems. *Proc. of the 1991 IEEE Conference on Robotics and Automation*, Sacramento, California, pp. 818-823.

[110] Maes, P., 1994, Modelling adaptive autonomous agents, *Artificial Life*, Vol.1, No.1-2, pp.135-162.

[111] Maimon, O.Z., 1987, Real-time Operational Control of Flexible Manufacturing Systems, *Journal of Manufacturing Systems*, Vol. 17, No. 6, pp. 125-136.

[112] Matsuo, H., Shang, J.S., and Sullivan, R.S., 1989, A knowledge-based system for stacker crane control in a manufacturing environment. *IEEE Trans. on Syst. Man and Cybernetics*, Vol. 19, No. 5, 1989.

[113] Matsuura, H., and Tsubone, H., 1986, A comparison of centralized and decentralized control rules in push type production ordering systems. *European Journal of Operational Research*, No. 25, pp. 272-280.

[114] Meseguer, P., 1989, Constraint satisfaction problems: an overview, *AI Communications*, Vol.2, No.1, March 1989, pp. 3-17.

[115] Milne, R., Ghallab, N.M., Trave-Massuyes, L., Bousson, K., Dousson, C., Quevedo, J., Aguilar, J. and Guasch, A., 1994, TIGER: real-time situation assesment of dynamic systems. *Intelligent Systems Engineering*, Spec Issue on the Second Int. Conference ISE'94, Hamburg, Germany, IEE - BSC Publ., London , 1994, pp.1-22.

[116] Minas, M., 1992, Uberwachung technischer Prozesse mit Zeit constraintnetzen, Arbeitsberichte des Institutes fur Mathematische Maschinen und Datenverarbeitung (*Ph.D. Dissertation*) Bd.25, Nr.3, University Erlangen-Nurnberg, May 1992.

[117] Minton, S., Johnston, M.D., Philips, A.B. and Laird, Ph., 1992, Minimising conflicts: a heuristic repair method for constraint satisfaction and scheduling problems, *Artificial Intelligence,* Vol. 58, 1992, pp.161-205.

[118] Mitchell, T.M., 1990, Becoming increasingly reactive, *Proceedings of the 1990 AAAI Conference*, Boston, pp.459-467.

[119] Miyashita, K., and Sycara, K., 1994, Adaptive Case-based Control of Schedule Revision. In: M. Zweben and .S. Fox (Eds.), *Intelligent Scheduling*, Morgan Kaufmann, San Francisco, California, 1994, pp.291-307.

[120] Monostori, L., 1996, From pattern recognition techniques through artificial neural networks to hybrid AI solutions in manufacturing. *Proc. of the 1996 Japan-USA Symposium on Flexible Automation*, Boston, Massachusetts, USA, July 1996, Volume 2, pp.1453-1460.

[121] Monostori, L., Markus, A., Van Brussel, H., and Westkamper, E., 1996, Machine Learning Approaches to Manufacturing. *Annals of the CIRP*, Vol. 45/2, pp.675-712.

[122] Monostori, L., Szelke, E., and Kadar, B., 1997, Management of changes and disturbances in manufacturing systems. In: *Proc. for the IFAC Workshop on Manufacturing Systems: Modelling, Management and Control - MIM'97*, February 3-5, 1997, Vienna, Austria, Vienna University of Technology, IFAC, 27-38.

[123] Montazeri, M., and van Wassenhove, L.N., 1990, Analysis of scheduling rules for a Flexible Manufacturing System. *Int. Journal of Production Research*, Vol. 28, No. 4, pp.785-802.

[124] Mulkens, H., 1994, Revisiting the Johnson algorithm for flow-shop scheduling with genetic algorithms. Proc. of IFIP Workshop on *Knowledge Based Reactive Scheduling*, E. Szelke and R.Kerr(Eds.), IFIP/ Elsevier (North Holland), Amsterdam, 1994, pp.69-80.

[125] Nii, H.P., Blackboard systems, *AI Magazine*, Vol. 7. No. 2 and 3, 1986, pp.38-53 and pp. 82-106.

[126] Noubissi, J.F., B. Beldjilali, D. Trentesaux, C. Tahon, 1994, Modelling with coloured Petri nets and simulation studies of a dynamic and distributed management system for a manufacturing cell. *Int. Journal of Computer Integrated Manufacturing*, Vol. 7, No. 6, 1994, pp. 323-339.

[127] O'Hare, G.M.P., 1990, Designing Intelligent Manufacturing Systems: A Distributed Artificial Intelligence Approach. In: Intelligent Manufacturing Systems-IMS'89 (Special Issue) *Computers in Industry*, Elsevier, Vol.15, (1990), pp.17-25.

[128] O'Hare, G.M.P. and Jennings, N.R. (Eds.), 1996, *Foundations of Distributed Artificial Intelligence*, John Wiley and Sons, New York.

[129] Oren, T.I., and Zeigler, B.P., 1979, Concepts for Advanced Simulation Methodologies. *Simulation*, Vol. 32, No. 3, pp. 69-82.

[130] Oren, T.I., and Zeigler, B.P., 1987, Artificial Intelligence in modeling and simulation: Directions to explore. *Simulation*, Vol. 48, No. 4, pp. 131-134.

[131] Ow, P.S., and Smith, S.F., 1988, Viewing Scheduling as an Oppportunistic Problem-Solving Process. *Annals of Operations Research*, Vol. 12, pp.85-108.

[132] Ow, P.S., Smith S.F. and Howie, R., 1988, A co-operative scheduling system, *Expert Systems and Intelligent Manufacturing*, Ed.: M.D. Oliff, Elsevier, 1988, Amsterdam, pp. 43-55.

[133] Owen, J.V. and Sprow, E.E., 1994, Shop-floor' 94 - The challenge of change, *Manufacturing Engineering*, Vol.12, 1994, pp. 33-46.

[134] Paul, C.J., Holloway, L.E., Yan, D., Strosnider, J.K., and Krogh, B.H., 1992, An Intelligent Reactive Monitoring and Scheduling System. *IEEE Control System Mag. (USA)*, Vol.12, No.3, June 1992, pp.78-86.

[135] Parunak, H.V.D., 1987, Manufacturing experience with the contract-net. In: *Distributed Artificial Intelligence*, (M.N. Huhns, ed.), Morgan Kaufmann, San Mateo, CA/Pitman, London, pp.285-310.

[136] Pegden, C.D., *Introduction to SIMAN*, System Modeling Corporation, Pennsylvania, 1984.

[137] Perkins, J.R. and Kumar, R.R., 1989, Stable, distributed, real-time scheduling of FMS, FAS and disassemble systems, *IEEE Trans. on Automatic Control*, Vol. 34, No. 2, 1989, pp. 139-148.

[138] Peterson, J.L., 1981, *Petri Net Theory and the Modeling of Systems*, Prentice-Hall, Englewood Cliffs, NJ, 1981, 290 pp.

[139] Pnueli, A., 1986, Specification and development of reactive systems - Invited paper for *World Congress of IFIP, Dublin, Ireland, In: Information Processing 86*, H.-J. Kugler (Ed.), Elsevier Sci. Publ. (North-Holland), IFIP, 1986, pp. 845-858.

[140] Prosser, P., 1989, A reactive scheduling agent, *Proceedings 11th Int. Joint Conference on AI*, Detroit, USA, 1989, pp. 1004-1009.

[141] Prosser, P., 1993, Scheduling as a constraint satisfaction problem: theory and practice, In: *Scheduling of Production Processes*, Eds.: J. Dorn and K.A. Froeschl, Ellis-Horwood, Vienna, 1993, pp. 22-30.

[142] Quinlan, J.R., *C4.5 Programs for machine learning*, Morgan Kaufmann Publishers, San Mateo, California, 1993.

[143] Rabelo, L.C., Yih, Y. and Jones, A., 1995, Knowledge Based Reactive Scheduling using the integration of Neural Networks, Genetic Algorithms and Machine Learning, In: *Proc. of the IFIP Int. Working Conference on 'Managing Concurrent Manufacturing to Improve Industrial Performance'*, Sept 11-15, 1995, Seattle, USA, Washington Univ. pp.444-455.

[144] Rabelo, R.J., and Camarinha-Matos, L.M., 1994, Negotiation in multi-agent based dynamic scheduling. *Robotics & Computer-Integrated Manufacturing*, Vol.11, No. 4, pp.303-309.

[145] Rabelo, R.J., and Camrinha-Matos, L.M., 1996, HOLOS: A methodology for deriving scheduling systems. In: *IEEE/ECLA/IFIP Int. Conference BASYS'95 on Architectures and Design Methods for Balanced Automation Systems*, July 1995, Victoria, Brasil, Chapman & Hall, London, 1996.

[146] Reynolds, D., and Cartwright, Ch., 1989, Case study in knowledge based fault diagnosis and control. In: *Knowledge-Based System Diagnosis, Supervision and Control*, Ed.: S.G. Tzafestas, Plenum Press, New York, pp. 153-165.

[147] Rodd, M., Verbruggen, H., Krijgsman, A., 1992, Artificial Intelligence on Real-Time Control, *Engineering Applications of Artificial Intelligence*, Vol. 5, No.5, pp. 385-399.

[148] Roy, U. and Zhang, X., 1996, A heuristic approach to n/m job shop scheduling: fuzzy dynamic scheduling algorithms, *Int. Journal of Production Planning & Control*, Vol. 7, No. 3, pp. 299-311.

[149] Sabuncuoglu, I., and Gurgun, B., 1996, A neural network model for scheduling problems. *European Journal of Operational Research*, Vol. 93, (1996), pp. 288-299.

[150] Sadeh, N., 1994, Micro-Opportunistic Scheduling: The Micro-Boss Factory Scheduler. In: *Intelligent Scheduling*. Eds.: M. Zweben and M.S. Fox, Morgan Kaufmann, San Francisco, 1994, pp. 99-136.

[151] Sadeh, N. and Fox M.S., 1990, Variable and value ordering heuristics for activity based job-shop, *Proc. of the Fourth Int. Conference on Expert Systems in Production and Operation Management*, Hilton Head Island, USA, 1990, pp.134-144.

[152] Sadeh, N. and Fox M.S., 1991, Micro-versa Macro-opportunistic Scheduling. *Proc. CAPE'91 IFIP Conf. on Computer Applications in Production and Engineering,* Elsevier (North -Holland) Amsterdam, pp.651-658.

[153] Sarin, S.C., and Salgame, R., 1989, A knowledge-based system approach to dynamic scheduling. In: Kusiak, A., (ed.), *Knowledge-Based Systems in Manufacturing*, Taylor & Francis, London, pp. 173-203.

[154] Sathi, A., and Fox, M.S., 1989, Constraint-directed negotiation of resource reallocation. In: *Distributed Artificial Intelligence* 2, (L. Gasser and M.N. Huhns, eds.), Morgan Kaufmann, Los Altos, CA/Pitman, London, pp.163-193.

[155] Schmidt, G., 1994, How to apply fuzzy logic to reactive scheduling. In: E. Szelke and R.M. Kerr (Eds.), *Knowledge Based Reactive Scheduling*, Elsevier (North-Holland), Amsterdam, 1994, pp.57-67.

[156] Schoppers, M., 1991, Real-time knowledge-based control systems. *Communications of the ACM*, Vol. 34, pp.27-30.

[157] Sen, A.K., Bagchi, A., and Ramaswamy, R., 1996, Searching Graphs with A*: Application to Job Sequencing, *IEEE Transactions on SMC - Part A: Systems and Humans*, Vol 26, No. 1, January, 1996, pp.168-173.

[158] Sidhu, S., Gupta, S., and Vlach, F., 1996, Issues in the design and implementation of intelligent scheduling systems, *Intelligent Manufacturing Processes*, 1988, pp.375-379.

[159] Shaw M.J., and Fox, M.S., 1993, Distributed Artificial Intelligence for Group Decision Support - Integration of Problem solving, Coordination and Learning. *Decision Support Systems*, 9, (1993), pp.349-367.

[160] Shukla, C.S., and Chen F.F., 1996, The state of the art in intelligent real-time FMS control: a comprehensive survey. *J. of Intelligent Manufacturing*, No. 7, pp.441-455.

[161] Smith, R.G., 1980, The contract net protocol: High-level communication and control in a distributed problem solver. *IEEE Trans. Comput.*, Vol.C-29, Dec. 1980, pp.1104-1113.

[162] Smith, R.G., and Davis, R., 1981, Frameworks for co-operation in distributed problem solving, *IEEE Trans. on Systems, Man and Cybernetics*, Vol. 11, No.1, pp.

[163] Smith, S.F.,1992, Knowledge-based production management: approaches, results and prospects. *Int. Journal of Production Planning & Control*, Vol. 3, No. 4, pp. 350-380.

[164] Smith, S.F.,1994, OPIS: A Methodology and Architecture for Reactive Scheduling. In: *Intelligent Scheduling*, (Eds.: M. Zweben and M.S. Fox), Morgan Kaufmann Publishers, San Francisco, California, 1994, pp. 29- 66.

[165] Smith, S. F., and Lassila, O., 1994, Configurable Systems for Reactive Production Mangement. In: *Knowledge Based Reactive Scheduling*, Eds.: E. Szelke and R.M. Kerr, Elsevier Science Publishers (North- Holland), Amsterdam, pp.93-106.

[166] Starkwether, T., Whitney, T., and Cookson, B., 1992, A Genetic Algorithm for Scheduling with Resource Consumption. *Proc. of the Joint German/US Conference on Operations Research in Production Planning and Control*, Springer-Verlag, Berlin, pp. 567-583.

[167] Stoffel, K., Law, I., and Hirsbrunner, B., 1993, Fuzzy logic controlled dynamic allocation system. In: B. Hirsbrunner, M.Courant and M. Aguilar (eds.), *Parallelism and Artificial Intelligence*, Series in Computer Science, University of Fribourg, Fribourg, 1993.

[168] Sullivan, J.W. and Tyler, Sh. W., 1991, *Intelligent User Interfaces*, ACM Press.

[169] Sutton,R., 1990, First results with DYNA, an integrated architecture for learning, planning and reacting. *Proc. of the AAAI Spring Symposium on Planning in Uncertain, unpredictable, or changing environment*, (AAAI, March, 1990), pp.136-140.

[170] Szelke, E., and Kerr, R.M., 1994, Knowledge Based Reactive Scheduling (Invited paper), *Int. Journal of Production Planning and Control*, Vol.5, March-April, 1994, pp.124-145.

[171] Szelke, E., and Markus, G., 1994, Reactive scheduling - An intelligent supervisor function, *Proceedings First IFIP Workshop on 'Knowledge Based Reactive Scheduling'*, Eds.: E. Szelke and R.M. Kerr, Elsevier (North-Holland), Amsterdam, 1994, pp.125-145.

[172] Szelke, E., and Markus, G., 1997, A learning reactive scheduler using CBR/L. *Computers in Industry* 33 (1997), pp. 31-46.

[173] Szelke, E., and Markus, G., 1995, A blackboard based perspective of reactive scheduling. In: *Artificial Intelligence in Reactive Scheduling*, R. Kerr and E. Szelke (Eds.), Chapman&Hall (IFIP), London, 1995, pp.60-77.

[174] Szelke, E., and Monostori, L., 1995, Reactive and proactive scheduling with learning in reactive operation management. *Proc. of the IFIP WG5.7 Int. Working Conference on 'Managing Concurrent Manufacturing to Improve Industrial Performance*, September 1995, Seattle, Washington University, IFIP Transactions, pp.456-483.

[175] Sycara, K.P., and Miyashita, K., 1994, Adaptive Schedule Repair, *Proceedings First IFIP Workshop on 'Knowledge Based Reactive Scheduling'*, Eds.: E. Szelke and R.M. Kerr, Elsevier (North-Holland), Amsterdam, 1994, pp.107-123.

[176] Sycara, K.P., Roth, S., Sadeh, N., and Fox, M., 1991, Resource Allocation in Distributed Factory Scheduling, *IEEE Expert Intelligent Systems and their Applications*, February 1991, pp.29-40.

[177] Sycara K., and Miyashita, K., 1993, Schedule Repair through Case-based Reasoning. Technical Report, The Robotics Institute, Carnegie Mellon University, September 1993.

[178] Sycara, K., Pannu, A, Williamson, M., Zeng, D., and Decker, K., 1996, Distributed Intelligent Agents, *IEEE Expert*, 1996, pp.36-45.

[179] Syswerda,G., 1991, Schedule Optimization Using Genetic Algorithms. *Handbook of Genetic Algorithms.* Ed.: L. Davis, Van Nostrand Reinhold, New York, 1991.

[180] Takatori, N., and Kakazu, Y., 1996, Dynamic Jobshop Scheduling by Hopfield Neural Networks. In: *Proc. of the Japan/USA Symposium on Flexible Automation*, Vol. 2, ASME 1996, pp.1419-1422.

[181] Tawegoum, R., Castelain, E., and Gentina, J.C., 1994, Dynamic Operations Control in Flexible Manufacturing Systems (FMS). *Proc. of the IFAC/IFIP/FORS Conf. on Intelligent Manufacturing Systems, IMS'94*, June 1994, Vienna, Austria, pp.285-290.

[182] Tchako, J.F.N., Beldjilali, B, Trentesaux, D., and Tahon, C., 1994, Modelling with coloured timed Petri nets and simulation of a dynamic and distributed mangement system for a manufacturing cell. *Int. J. Computer Integrated Manuf.*, Vol.7, No. 6, pp.323-339.

[183] Tsang, E. 1993, *Foundations of Constraints Satisfaction.* In: Series for Computation in Cognitive Science, Academic Press, Harcourt Brace & Company Publishers, 1993.

[184] Tzafestas, S.G. (Ed.), 1989, *Knowledge-Based Diagnosis, Supervision, and Control.* In: Applied Information Technology series, Plenum Press, New York, 1989.

[185] Uma, G., Prasad, B.E., and Kumari, O.N., 1993, Distributed intelligent systems: issues, perspectives and approaches. *Knowledge Based Systems*, Vol 6, No. 2, pp. 77-86.

[186] Vaithyanathan, S., and Ignizo, J.P., 1992, Stochastic neural network for resource constrained scheduling, *Computers and Operations Research*, 19, pp. 241-254.

[187] Valette, R., Cardoso, J., and Dubois, D., 1989, Monitoring manufacturing systems by means of Petri nets with imprecise markings. *IEEE-TH0282-4/89*, pp.233-239.

[188] Vancza, J. and Bertok, P.,1989, Knowledge-based tool for manufacturing control: a general view. In: *Knowledge-Based Diagnosis, Supervision, and Control.* Ed.: Tzafestas, S.G., Applied Information Technology series, Plenum Press, New York, 1989, pp. 231-246.

[189] Van der Aalst, W.M.P., 1994, Putting high-level Petri nets to work in industry. *Computers in Industry*, Vol. 25, pp.45-54.

[190] Van der Aalst, A., 1989, Expert monitoring and control of failure-prone flexible manufacturing shop. *Advanced Manufacturing Engineering*, Vol. 1, April 1989, pp.147-154.

[191] Villa, A., 1991, *Hybrid Control Systems in Manufacturing*. Gordon and Breach Science Publishers, London.

[192] Villa, A., 1994, Decentralized production scheduling by a network of local intelligent controllers. In: *Proc. 10th Int. Conference on Computer Aided Production Engineering*.

[193] Villa, A., Mosca, R., and Murari, G., 1986, Expert Control Theory: A key for solving production planning and control problems in flexible manufacturing. *IEEE - CH2282-2/86*, pp. 466-471.

[194] Wiendahl, H.-P., Garlichs, R., 1994, Decentral Production Scheduling of Assembly Systems with Genetic Algorithm, *Annals of the CIRP*, Vol.43/1, 1994, pp.389-395.

[195] Wiendahl, H-P., Ludwig, E., and Ullmann, W., 1994, Monitoring and Diagnosis Systems: New Components of Production Planning and Control - Methodologies, Applications and Experience. *Int. Trans. Opl. Res.,(Elsevier Science Ltd.)*, Vol. 1, No. 1, pp. 95-106.

[196] Wiendahl, H.-P., and Scholtissek, P., 1994, Management and Control of Complexity in Manufacturing. *Annals of the CIRP*, Vol.43/2, pp.533-540.

[197] Willems, T.M., 1994, Neural networks for job-shop scheduling. *Control Engineering Practice*, Vol. 2, No. 1, pp.31-39.

[198] Willems, T.M., and Brandts, L.E.M.W., 1995, Implementing heuristics as an optimization criterion in neural networks for job-shop scheduling. *J. of Intelligent Manufacturing,* Vol.6, pp.377-388.

[199] Willson, R.G., and Krogh, B.H., 1990, Petri net tools for the specification and analysis of discrete controllers. *IEEE Transactions on Software Engineering*, Vol. 16, pp.39-50.

[200] Wu, H.-H. and Li, R.-K., 1996, A methodology for schedule compression. *Int. Journal of Production Planning & Control*, Taylor & Francis, London, Vol. 7, No. 4, pp.407-417.

[201] Wu, H.-H. and Li, R.-K., 1994, A new rescheduling method for computer-based schedule systems. *Int. Journal of Production Research.*

[202] Wu, H.-J., and Joshi, S.B., 1994, Error recovery in MPSG-based controllers to shop floor control. *Proc. IEEE Int. Conference on Robotics and Automation*, San Diego, California, 8-13 May, pp.1374-1379.

[203] Wu, H.-J., and Wysk, R.A., 1989, An application of discrete event simulation to the on-line control and scheduling in flexible manufacturing. *Int. Journal of Production Research*, Vol.27, No. 9, pp.1603-1623.

[204] Ye, N., Zhao, B., and Salvendy, G., 1993, Naural-Networks-Aided Fault Diagnosis in Supervisory Control of Advanced Manufacturing Systems. *Int. Journal of Advanced Manufacturing Technology*, Springer Verlag, London, Vol. 8, pp.200-209

[205] Yih, Y., 1990, Trace-Driven Knowledge Acquisition (TDKA) for Rule-Based Real-Time Scheduling Systems. *Journal of Intelligent Manufacturing*, Vol. 1, No. 4, pp.217-230.

[206] Yih, Y., 1992, Learning real-time scheduling rules from optimal policy of semi-Markov decision processes. *Int. J. Computer Integrated Manufacturing*, Vol.5, No.3, pp.171-181.

[207] Yuan, Y., and Wu, Z., 1991, Algorithm of fuzzy dynamic programming in AGV scheduling. *Proc. Int. Conference on Computer Integrated Manufacturing, ICCIM'91, 'Manufacturing Enterprises of the 21st Century'*, World Scientific Publ., Singapore, pp.405-408.

[208] Zadeh, L.A., and Kacprzyk, J., 1992, *Fuzzy Logic for the Management of Uncertainty*, John Wiley and Sons, New York, 1992.

[209] Zeigler, B.P., 1984, *Multi-Faceted Modelling and Discrete Event Simulation*, Academic Press, New York.

[210] Zeigler, B.P., 1989, DEVS representation of dynamical systems: event-based intelligent control. *Proc. of the IEEE*, Vol. 77, pp.72-80.

[211] Ziegler, B.P., 1990, *Object Oriented Simulation with Hierarchical, Modular Models: Intelligent Agents and Endomorphic Systems*, Academic Press, New York.

[212] Zeigler, B.P., Cho, T.H., and Rozenbilt, J.W., 1996, Knowledge-Based Simulation Environment for Hierarchical Flexible Manufacturing, *IEEE Trans. on Systems, Man and Cybernetics-Part A: Systems and Humans*, Vol. 26, No. 1, January 1996, pp. 81-89.

[213] Zweben, M. and Fox, M. S. 1994. *Intelligent Scheduling*. Morgan Kaufmann Publishers, San Francisco, California, 1994.

5

Simulation within CAD-Environment

Peter Kopacek[1]
Gernot Kronreif[2]
Tomaž Perme[3]

5.1 Introduction

In consideration of the present market situation, there is an increasing demand on high quality products to a reasonable price. To meet these requirements, appropriate manufacturing technologies have to be defined and manufacturing resources have to be utilised in a sufficient way. Market's imponderableness however, asks for a higher flexibility in production. Small and medium sized companies must therefore have high production flexibility to be able to meet the rapidly changing market situation, to meet the delivery dates, and to provide for the increasing demands regarding quality. Especially in automated assembly and disassembly systems, where the technology requires cost intensive resources (robots, end effectors and automated transport devices), a reliable planning of production systems has to be considered. Consequently, the planning of assembly systems as well as the generation of optimal assembly plans are of great importance.

During the planning phase of a manufacturing system both, performance and main costs of the system have to be determined. With different analysing tools the designed performance of the system has to be obtained and, if it doesn't match with the demands, some necessary corrections have to be done (Fig. 5.1). Analysis is also needed to obtain the behaviour of an existing manufacturing system – controlled or scheduled by different algorithms – before their implementation on a real system. Resulting from the decreasing life-time of certain products, another important factor – besides the reduction of planning costs – is the saving of time in the planning stage. Conventional planning techniques are not able to solve these problems in a sufficient way. When looking for innovative and economic solutions, instruments which show the weaknesses in a highly integrated and dynamic process and which support the acquisition and realisation of solutions and their utilisation, have to be taken into

[1]Institute for Handling Devices and Robotics, Technical University of Vienna, Floragasse 7A, 1040 Vienna, Austria, E-mail: kopacek@ihrt1.ihrt.tuwien.ac.at
[2]Institute for Handling Devices and Robotics, Technical University of Vienna, Floragasse 7A, 1040 Vienna, Austria, E-mail: kronreif@ihrt1.ihrt.tuwien.ac.at
[3]Laboratory for Handling and Assembly Systems Automation, Faculty of Mechanical Engineering, University of Ljubljana, Aškerčeva 6, 61000 Ljubljana, Slovenia, E-mail: tomaz.perme@fs.uni-lj.si

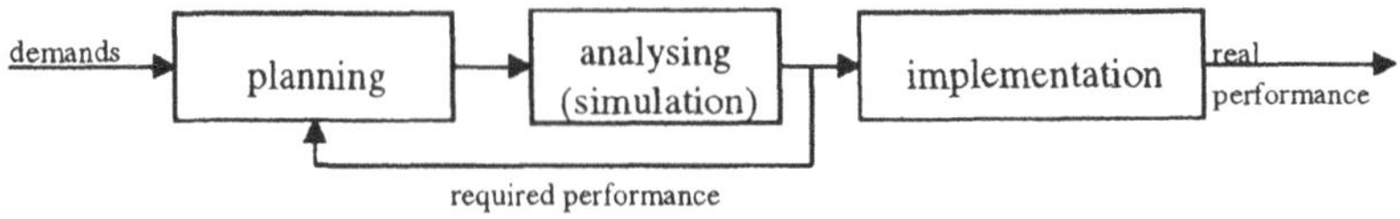

FIGURE 5.1. The role of simulation

consideration. Under this point of view **simulation** seems to be one of the most appropriate techniques.

Simulation is the imitation of the real plant in a computer model to deserve the dynamic behaviour under several variants of load and eventual breakdowns. Based upon the costly and time-consuming experiments which result from the real process, simulation offers a lot of help in the development and modification of complex systems. The financial benefit of simulation-based planning is out of the question nowadays. The utilisation of simulation technique permits a reduction of time spent in planning and of planning risk as well as an improvement in the quality of planning.

A positive side effect of the modelling is its extraordinarily high insight into the system. The plant is look at in a different way – from a different point of view. The formal description of the system, as it is necessary for the construction of a computer model, acts like a check-list. In many cases, eventual errors are already discovered at this stage before the model itself is built up and existing but hidden reserves of the plant are recognised. Simulation in connection with the animation of the manufacturing or the assembly process improve the communication between all the factors taking part in the process of planning.

Initially a tool for providing cover against planning risks, simulation now can be utilised in every particular stage of planning and realising the system as well as for control of the manufacturing process. Since of the unquestionable advantages of this tool, computer aided simulation must **and will** gain more importance in future.

Being well known as a planning tool for a considerable period of time, simulation mostly is not 'the first choice' in order to find solution to the problems mentioned above. Some possible reasons for this unsatisfactory acceptance might be:

- First and second generations simulation software and the poor support for model development led to the (wrong) opinion of an extremely costly and highly complicated tool.
- Poor organisational connection to departments and working groups.
- Bad interfaces to existing data bases and data processing systems.

Considering this, an improvement of simulation systems in respect to flexibility, interfaces to particular (hardware-) components of the manufacturing systems, user-orientation for modelling and experimentation tasks as well as complete integration of simulation technique into the whole planning cycle will be necessary.

5.1.1 SIMULATION AND CAD

Above all, integration of the two essential CA-techniques for design and layout planning – Computer Aided Design (CAD) and (Computer Aided) Simulation – seems to be advisable. Generally the CAD drawing of the plant layout contains many data used by the simulation system, as there are information concerning the physical elements of the plant (e.g. machines), information to the flow of materials, transport routes, and others (Fig. 5.2). Further data could be available in PPS (production planning systems) and/or in plant control systems. When opening the access to data stored in other CA-systems and using them for (partially) automated generation of a simulation model, the efforts and costs necessary for a simulation study can be decreased dramatically [C.92].

Especially for simulation in robotics the CAD system serves for generation of the geometrical models of robots, peripheral devices, and of all parts to be manipulated. Using these CAD-data simulation in robotics requires some additional features, like definition of the kinematics model of the devices, 'off-line' collision detection and avoidance, path- and assembly planning features, definition of the dynamic robot model. Usually these functionality is not available in CAD systems, which again leads to the necessity of an appropriate 'coalition' between CAD and simulation functions.

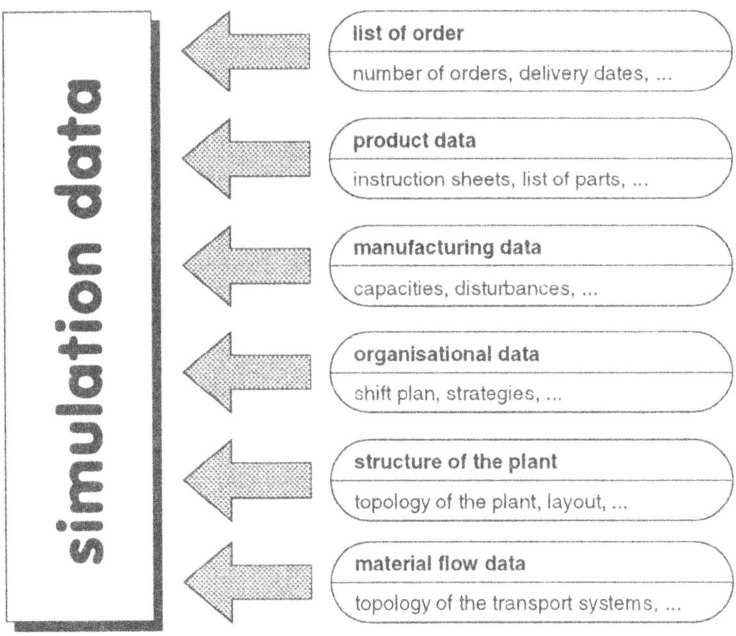

FIGURE 5.2. Input data for simulation.

There are different ways to form a combination between CAD functionality and simulation features:

- *CAD-simulation interface*
  The goal of such a CAD-simulation interface is to use the functionality of each tool during the appropriate planning stage. Most of known approaches of a combination between CAD and simulation systems are limited to the export of CAD-layout data for providing a more realistic graphical representation of the simulation model. Hence, the CAD layout is being reduced to a static animation background – the information included in the CAD layout, like transport routes, gets lost. For simulation in robotics, a CAD-simulation interface usually is used to import geometrical models of the cell components in order to provide a realistic graphical cell representation – data exchange mostly is done by using one of the standardised formats, like DXF, IGES, STEP.
- *CAD functions within simulation environment*
  In general there are some CAD functions available in most simulation systems for design of the animation layout (i.e. animation background and facilities). Making a simulation/animation model therefore consists of two steps: drawing of the animation layout and definition of links between the simulation model and the particular elements of the animation layout. In case of simulation systems for robotics, there are many commercial available packages (GRASP, ROBCAD, ROBSIM) which are providing more or less basic drawing functionality for definition of cell components, like robots, feeders, transport devices and other auxiliary devices [G.94].
- *Simulation within CAD-environment*
  The CAD-system serves as a carrier for an embedded sub-system with basic simulation features. As it can be seen in figure 5.1, simulation based planning is a highly iterative procedure. After arrangement of all cell components and programming of the devices, this layout is being checked by means of a set of simulation experiments. Analysis of the simulation results may cause a rearrangement of components, which again may lead to changes for the (CAD) cell layout. Therefore, a combination of CAD and simulation functionality into one software system seems to be very advisable. This type of 'interface' is used very often in simulation of robot-equipped systems. As already mentioned, the functionality of the used CAD system has to be extended with simulation features. As a consequence, the CAD system must have an appropriate programming interface which allows programming of new application modules, like modelling of the kinematics structure of manipulators, definition of links between geometrical elements and particular joints of the manipulator, grouping drawing elements (point, line, curve) and data sets into elements of the simulation model, simulation event control, and others.

The two approaches for a CAD-simulation 'combination' presented in this paper are both influenced by the special needs of small and medium sized companies. As a consequence, one of the most important features always kept in mind during the program development was to find a real 'low-cost' solution – without having the need of expensive hard- and software and/or costly operator training for running these software tools.

As one result of this demand, CAD package AutoCAD is being used as a system platform. This system – maybe the most common CAD software used in architecture, electronics, and engineering – has its advantage in supporting various hardware platforms and operating systems (DOS, WINDOWS, WINDOWS NT, HP-UX, SUN, etc.). Some other features recommending AutoCAD for our applications are:

- Graphical user interface with pull down menus, dialogue boxes, accelerator keys, icon toolbar and floating toolbox. Menus can be customised to suit user needs.
- AutoCAD Development System **ADS** supports a programming interface to C or C++ code for development of custom AutoCAD applications. Furthermore there are additional interfaces to other graphic standards and database systems (AutoCAD SQL Extension **ASE**).
- Performance: fast redraw, pan and zoom speed; features which are important especially for large drawings.

5.2 Simulation System LASIMCO

This chapter points out some problems in the planning of assembly systems which refer to the applicability and efficiency of the simulation technique. It also lists some requirements that simulation systems should satisfy. A concept which meets these requirements is introduced and the applied theory used in the simulation system LASIMCO is explained. After the description of LASIMCO, two examples which show its applicability are introduced.

5.2.1 Formulation of Requirements

The primary aim of assembly system planning is to define the most appropriate system regarding the available resources and set requirements. During the designing or redesigning of assembly systems, different solutions are found and evaluated and then the most appropriate one is selected. For complex systems that assembly systems are, evaluation cannot be done analytically – simulation is the tool to choose.

A model of the system which represents its structural properties and its analysis which gives information about its dynamic behaviour create the basis for the control process in the operation phase. It seems to be useful to apply the same model both for control strategy development and for evaluation during selection, because this reduces simulation time and costs. An applied modelling technique must show the system in more or less detail and, in addition, it has to be supported by certain analytical tools which can be used to solve some control problems and to validate the simulation model as well.

For experimentation, i.e. execution of simulation, appropriate presentation of results is very important. It can be performed in textual or graphical form. Some properties or states of the observed system, for example the temporary occupancy of assembly sites or pallet places on the conveyor belt, can be depicted more effectively if they are shown as 2D or even 3D graphics. This is very important in the phase of model verification and in the phase of experimentation when certain special situations have to be analysed. This deficiency can be successfully eliminated by introducing more realistic and accurate graphical presentation and animation. It usually leads to an enormous increase of computational demands and simulation therefore becomes slower.

The following requirements for the model can be inferred from the above problems:

- simulation systems must use the data generated during previous planing activities as efficiently as possible;
- there must be a possibility for the simulation system to be used as a part of the optimisation procedure;
- the simulation model needs to be able to show the observed system in different degrees of detail and analyse the system with analytical tools, and
- it must be able to show results by way of 2D or 3D graphics and animation, with execution speed not falling below acceptable limits.

5.2.2 Conceptual solution and applied theory

The conceptual solution includes answers regarding the above-mentioned requirements and can be formulated as follows:

- If CAD and simulation systems are integrated, the data contained in the layout can be automatically used for simulation. The layout specifies the spatial relationships between the components of an assembly system and gives their descriptions and structures. Components, such as assembly, transport, handling and other devices, can be shown by way of 2D or 3D graphics and used for animation.
- In selecting the optimal assembly system, minimisation of assembly costs should be the basic criterion. It can be calculated from the list of necessary equipment and the assembly system's capacity obtained by simulation. Figure 5.3 shows the position and flow of layout optimisation as part of the planning process.
- The assembly system can be viewed as a discrete event dynamic system. One of the most appropriate tools which can be used for modelling and analysing such a system is the Petri nets theory [RA90] [T.89] [SS94] [YP93]. It is a powerful modelling tool supported by analytical tools and can be used as the basis for simulation.
- The simulation system can be integrated within CAD, not just for simulation model generation and input data preparation, but also for the presentation of results. Results, which describe changes of moving components and other component properties, can be depicted as 2D or even 3D models. Through embodiment of the simulation system in CAD, animation can be performed using 2D or 3D models of components for a more realistic presentation of system behaviour.

Theory of Petri nets

The many fundamental concepts of Petri nets were defined by different researchers with different motivations and in different ways. A formal definition of a Petri net is given by [T.93] as a quintuple, $C = (P, T, F, W, M_0)$ where: $P = \{p_1, p_2, ...p_m\}$ is a finite set of places, $T = \{t_1, t_2, ...t_n\}$ is a finite set of transitions, $F \subseteq (P \times T) \cup (T \times P)$ is a set of arcs, $W : F \rightarrow 1, 2, 3, ...$ is a weight function, $M_0 : P \rightarrow \{1, 2, 3...\}$ is the initial marking and $P \cap T = 0$ and $P \cup T \neq 0$.

A Petri net is a directed, weighted, bipartite graph consisting of two types of nodes, called places and transitions. In graphical presentations, places are represented by circles and transitions by bars. Places are connected with transitions and transitions with places. The weight of arcs is a positive integer where a k-weighted arc means k parallel arcs. A marking assigns a non-negative integer to each state and is denoted by M; M is an $m \times 1$ column vector, where m is the number of places. If a non-negative integer k is assigned to place p, this place is marked with k tokens. In graphical presentations, tokens are represented by dots. According to firing rules, the analysis of Petri nets can be performed and the behavioural properties of the system can be investigated.

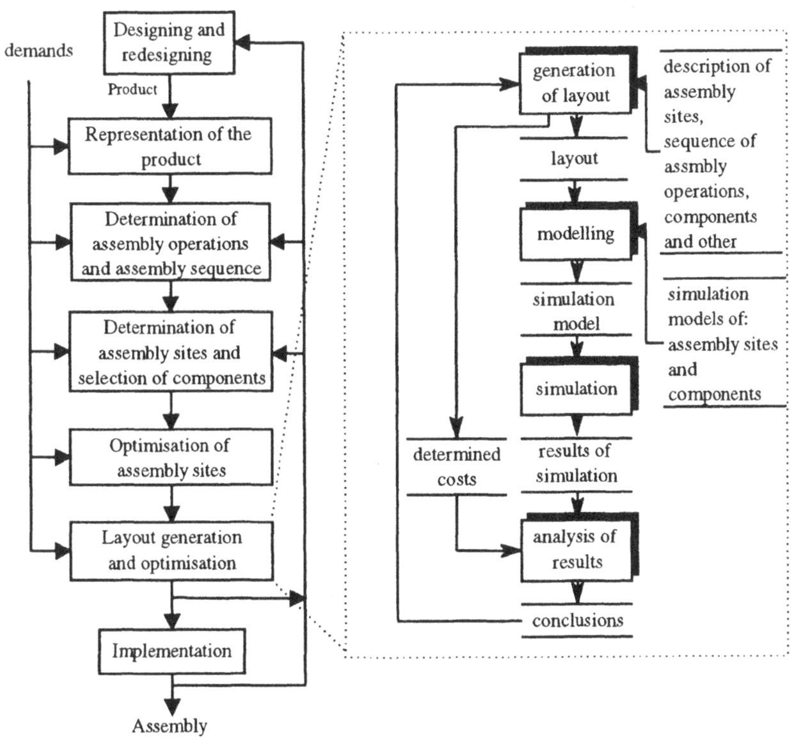

FIGURE 5.3. Flow of assembly planning with details of layout generation and simulation.

Modelling with Petri nets

The theory of Petri nets is very useful for the modelling of systems with characteristics such as concurrence or parallelism, deadlock and conflicting situations, asynchronous operations and sequences of discrete events which occur concurrently in the system. Assembly systems consist of components such as robots, grippers, manipulators, assembly devices, assembly tools, transport systems, computers and others, which are named resources. They are united with materials and information during activities required for assembly that is determined by a sequence of assembly operations to be accomplished. Figure 5.4 shows an example of an assembly system which consists of one manual and one automated assembly site connected with a pallet transport system. The system is shown by basic graphical models as well as a Petri net model. Both models are needed for simulation with the LASIMCO program, but the user needs to build only a CAD model, because a Petri net model is built automatically by the program.

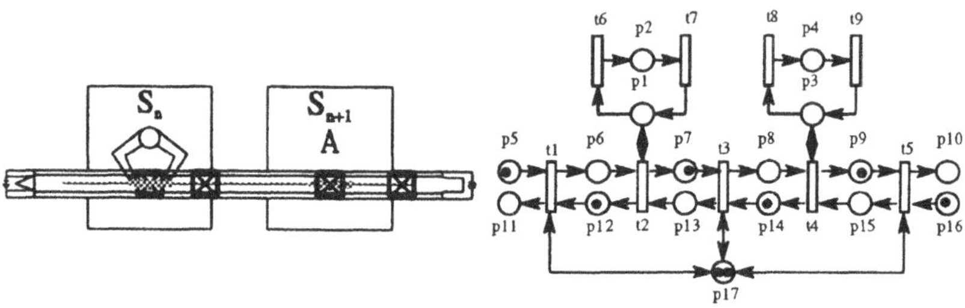

FIGURE 5.4. Assembly system presentation with basic graphical models in LASIMCO and Petri net model. Places p1 and p3 represent the ready states of the operator and the automated site, respectively. Places p2 and p4 represent stop and failure states of the operator and the automated site. Places from p5 to p11 represent control information on the states of sections of the conveyor and sites. Places from p11 to p16 represent the states of conveyor places, e.g. how many pallets there are on a certain part of the conveyor. Place p17 represents ready the state of the conveyor.

Analysis methods

There are two main analysis methods for Petri nets: reachability tree meth-od and matrix-equation approach. The applicability of the reachability tree method is limited for small nets due to the complexity of state-space explosion, but it can be used with benefit for parts of models. The matrix-equation method allows us to describe and analyse the dynamic behaviour of a Petri net through equations. The core of this method is an incidence matrix, an equivalent to the standard representation form, which allows the definition to be restated in vector and matrix terms. However, in many cases the matrix equation applies only to special subclasses of Petri nets or to special situations, because the solvability of equations is somehow limited by the non-deterministic nature inherent in Petri-net models and partly because solutions must be found as non-negative integers. For complex concurrent discrete event driven systems, a simulation technique is therefore still considered as the only possibility of obtaining objective results within a reasonable period of time. Petri nets based simulation is an imperative for analysing such systems because of the modelling power of Petri nets and because of direct implementation of Petri net models of observed systems in the simulation [T.93] [TD94].

Simulation with Petri nets

A matrix equation 5.1 used to develop a play algorithm for the execution of a Petri net serves as the basis for simulation. From the transition and firing rules, a sequence of firing vectors $\{u_1, u_2, ..., u_d\}$ can change their state from marking M_0 to marking M_d according to equation

$$M_d = M_0 + A^T \sum_{k=1}^{d} u_k = M_0 + A^t x, \tag{5.1}$$

where x is $n \times 1$ column vector called the firing count vector. The concurrence and duration of activities with certain extensions to an ordinary playing algorithm, which executes the Petri net in accordance with the transition and firing rules, are implemented for the simulation of Petri net models of systems characterised by mutual exclusion. For the simulation of Petri nets with transitions which represent events and activities with durations longer than zero, the theory of Petri nets is extended by introducing $d = \{d_1, d_2, ...d_j\}$. d is $n \times 1$ column vector of non-negative integers, where d_j denotes the time required to accomplish an activity represented by transition t_j. When time is introduced, the playing algorithm for the execution of a Petri net can perform the play in the manner described below.

Step 0. Petri net model is searched for enabled transitions. Simulation time is set to zero.

Step 1. All enabled and allowed transitions fire partially (enabled transitions are those whose input places are sufficiently marked; allowed transitions are enabled transitions which are in a conflict situation, but have firing priority over others; to fire partially means that the transition takes tokens from its input places).

Step 2. Untimed transitions fire completely. Other transitions are placed into a bag of partially fired timed transitions ('fire completely' means that a transition puts tokens in its output places; timed transitions have duration times which differ from zero).

Step 3. A Petri net is searched for enabled transitions. If there are existing enabled transitions, proceed with step 1.

Step 4. If there are no enabled transitions, timed transitions with shortest waiting times among the transitions in the bag fire completely. The simulation time increases, while waiting times of transitions in the bag decrease by this time. The procedure proceeds with step 2.

Step 5. If no enabled transitions are left, or if no partially fired transitions are left in the bag, simulation is completed.

5.2.3 Developed tools

The LASIMCO simulation program was developed for the planning of pallet assembly systems. It includes four modules that directly support planning activities through layout design and optimisation explained in the figure5.3.

These modules are:

- GenLay for layout generation;
- LaySim for simulation model generation;
- LasSim for execution of simulation and
- CosCal for the presentation of results and calculation of costs.

Layout generation

GenLay is a module that helps the designer to generate a layout of the pallet assembly system by putting a 2D representation of assembly stations and standard modules of the pallet transport system in place and linking them through different connections. Designing a layout with this module is like building a real system and it can be described in three steps:

The first step involves the preparation of a place and environment to start the building procedure. According to the description of assembly stations, all non-standard or new drawings and models can be generated and stored for further use. If there is an existing layout drawing of the works area showing all the limitations and obstacles (walls, pillars and already installed devices) that need to be considered, it can be imported in the drawing and certain areas can be marked as forbidden to cross or to place anything on them. In addition, some variables have to be set, which are used in the model and can be reset during the design procedure: a) types and dimensions of pallets used in the system, b) time unit, which defines the smallest time unit allowed as an input and c) space violation variable, which enables and disables a program for checking for space violation.

Second step. The assembly stations and transport devices that are part of the planned system are put in place. For each of them, a user-defined drawing or standard presentation is selected and an input form is filled with data describing its properties. There are two standard input forms (for the assembly station and for the transport system) with descriptions. The assembly station can be described by its dimensions and orientation, type of assembly process, pallet dimensions and orientation, costs, time needed for the execution of assembly operations, time between two failures and time needed for recovery. The elements of the transport system are described in the same way as assembly places; only assembly time is replaced by the velocity of the transport element. These properties, which do not cause any graphical changes, can be changed during the design procedure.

The third step is connection of transport system elements and assembly places according to the prescribed pallet flow. When an appropriate type of connection is chosen, two selected objects can be connected. In cases of junctions, certain decisions about the manner in which pallets should proceed can be defined by a simple control strategy embedded in a simulation model for connections.

All data defining the layout – and indirectly the simulation model – are part of an AutoCAD drawing. The simulation model can be generated when all objects are properly connected.

Simulation modell generation

LaySim is a computer program which automatically generates a simulation model of the system on the basis of data from layout drawing. The Petri net theory is applied for modelling. Each object in the layout has its own simulation model described and represented by an incidence matrix, list of times, priorities and initial marking.

The simulation models of assembly stations and elements of the pallet transport system are merged together with simulation models of connections into a consistent simulation model for the entire system. The outputs of this program are the system's incidence matrix, the time vector, initial marking and the vector of priorities. They are stored in an ASCII file which is used further as simulation model.

Execution of simulation

Experimentation with the simulation model can be divided to initialisation and execution of simulation. The parameters of the observed system, such as the number and place of pallets, observed properties and duration of simulation runs, have to be defined during the initialisation phase. The place and number of pallets can be defined by entering the desired number of pallets in the layout. The pallets can be entered even in assembly stations, so that the start-up period has less influence on simulation results. The properties of the system (e.g. the number of assembled products or use of a particular assembly station), which are observed and need to be recorded, can also be selected.

The **LasSim** simulation program executes a play algorithm described in the previous section. With the use of this program, experiments are repeatable, because the random number generator can be initialised and the sequence of random numbers can therefore be controlled. Each stochastic parameter has his own random number generator, so that the change of one parameter has no influence on sequences of random numbers for other parameters. The Chi-square method is implemented for the statistical processing of simulation results. The method gives a number of simulation runs that have to be conducted if one desires simulation results to be within a 95 % confidence interval. The program can perform simulation in tree different modes. In the first mode, the program executes simulation only once. In the second mode, the program runs simulation for a set number of times and in the third one the program runs simulation as many times as is needed for the results to satisfy the confidence interval. In the second and third modes, each simulation run is executed automatically, each time with a different initialisation seed of the random number generator.

The simulation program is written in the C language for DOS and ADS, so simulation can be executed either within the AutoCAD or stand-alone. The only difference between these two versions lies in the update of graphical presentation of pallet position in the layout (at stop time) using simulation within AutoCAD.

For stand-alone simulation, the simulation model and the initial state need to be prepared and stored in an ASCII file. With this DOS simulation program, simulation can be executed in shorter period of time, therefore a greater number of simulation runs or longer ones can be executed more efficiently. Because the simulation model and the initial state are stored in ASCII files, it is further possible to use the simulation program (translated) for other operating systems, even on faster computers and workstations.

Presentation of results and calculation of costs

CosCal analyses and presents the results of simulation and calculates the costs of assembly. For each observed place in a model, a time diagram of its marking can be generated. It represents the utilisation of buffer or assembly station capacities. The mean value and standard deviation of tokens in observed places can also be calculated. If an observed place in the model represents an assembly station at which finished assemblies are taken from the system, the number of passed tokens in that place equals the number of finished assemblies in the system, that is the expected system capacity based on the simulation.

Assembly costs per product are calculated from the pallet assembly system costs, accumulation time and expected system capacity. The total costs of a pallet assembly system are calculated automatically from the layout. Each assembly station and each module of a pallet transport system have fixed and variable costs, which are determined in the layout. Total costs are the sum of all fixed and variable costs multiplied by the total number of finished products. The accumulation time defines the life-time of an assembly system or product and is used with the expected system capacity in the calculation of the total number of finished products in the system.

5.2.4 EXAMPLES

The LASIMCO program was used to analyse the behaviour of an existing assembly system under different operating conditions (example 1) and to obtain data on the performance of the planned assembly line (example 2).

Example 1

An analysis of an existing pallet assembly system can be used as an example of the application of LASIMCO as a modelling and analytical tool. In the DOMEL company, the absorption part of a vacuum cleaner is assembled on a pallet transport assembly system. The system consists of six manual and six automatic assembly stations, a manipulator and an output station. Assemblies are transported using the pallet transport system. Figure 5.5 shows the assembly system layout generated by the layout generation part of LASIMCO, where all objects as well as the border are presented with real dimensions. The line around the system defines its spatial limits, while the arrows in the transport system indicate the type of connection and the direction of pallet flow.

The operating times of manual assembly operations are shown in Table 5.1; the times of automatic assembly sites, and manipulation and the output station are shorter than 5 s. The velocity of the pallet transport system is 16 m/min and transport times are calculated automatically from the given velocity and distances between connections. The disturbances caused by employees who need to leave due to personal needs and other organisational demands were used as variable operational parameters. Each employee uses 17 minutes per 8 hours for personal needs and 2 minutes per hour to supply the work station with assembly parts. All these times are obtained by statistical processing of shop floor measurements.

TABLE 5.1. Operating times of manual assembly operations

Manual assembly site	WS-1	WS-2	WS-3	WS-4	WS-5	WS-6
Duration of assembly	15.5 s	14.7 s	14.7 s	15.5 s	14.5 s	15.7 s

Four experiments have been conducted, where four alternatives have been examined:

A1 - operation with all disturbances;

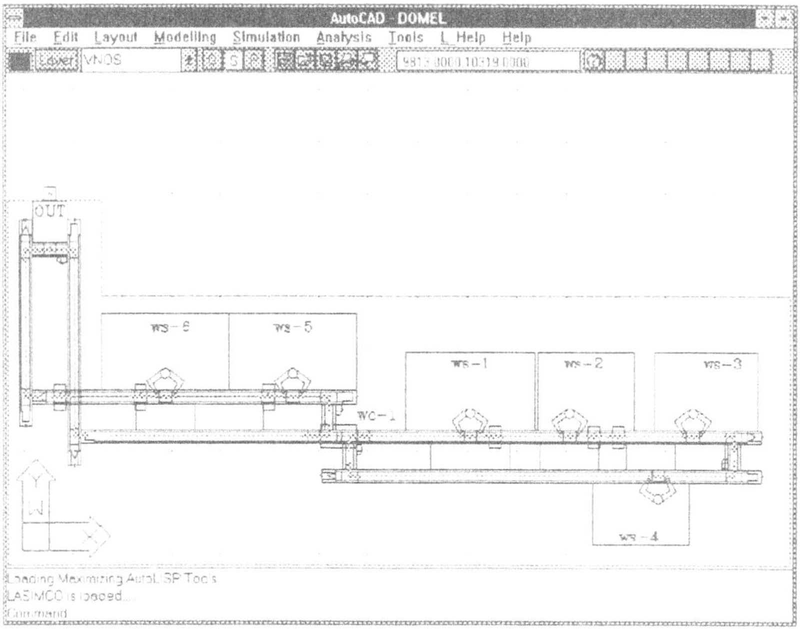

FIGURE 5.5. Layout of the analyzed system.

A2 - operation with all disturbances, but with unlimited buffer capacity ahead of manual assembly sites;

A3 - operation with an additional employee that supplies the assembly stations with assembly parts;

A4 - operation with an additional employee that supplies the assembly stations with assembly parts and takes over assembly, when other employees need to leave.

TABLE 5.2. Results of simulation and impact on output

Alternative	A1	A2	A3	A4
Troughput frequency	17.8 s	17.1 s	17.1 s	15.7 s
Increase of troughput	0 %	4.1 %	4.1 %	13.4 %

Based on these inputs, simulation models for each alternative were automatically generated, initialised and a sufficient number of simulation runs per alternative was executed, so that the results passed the 95 % confidence test. The duration of each simulation run was 7 hours and 30 minutes (duration of one shift), and the number of pallets passing the output station was the observed parameter. The results are presented in Table 5.2. They show that output increase for alternatives A2 and A3 is not significant; these alternatives therefore cannot be implemented either because of spatial limitation (A2) or because they cannot cover the costs for an additional employee (A3). The implementation of alternative A4 increases assembly line throughput by 13.4 %. If an investment in the assembly system of 750.000 money units, an accumulation time of 5 years and an annual salary per employee of 20.000 money units are taken into account, the costs of assembly for one product decrease by 4.8 %. Further details and results for this example are given in [TD94].

Example 2

LAMA, a company which produces metal fittings and assembles them in very large batches, plans and builds automated assembly lines for its own needs as well as for sale. Their assembly lines are mainly composed of various standardised modules designed and produced in the company. To be able to make the necessary changes and meet their customer's demands before the assembly lines are installed, they need to obtain data on line capacity in the design phase.

To be able to observe the impact of assembly part accuracy on the line throughput rate, the company had wished to install sensors on their operating assembly lines and monitor the assembly process. This was intended to provide information on the frequencies and causes of faults which caused individual assembly stations or the entire assembly line to stop. The obtained data were expected to be used later within the designing of new assembly lines.

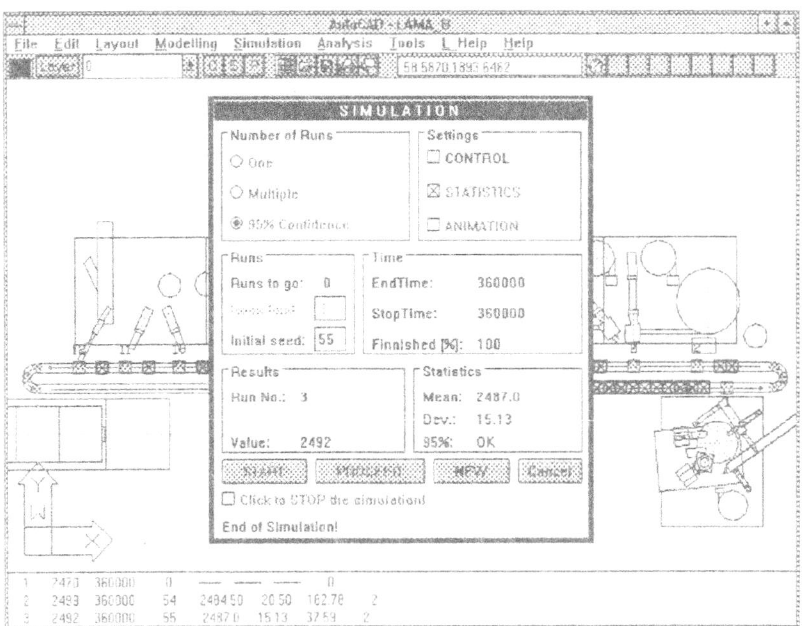

FIGURE 5.6. Layout of the assembly line LAMA BASIC.

In the first phase, the existing layout drawing was used as a template in positioning assembly sites and the pallet transport system (Fig. 5.6). The assembly sites defined the positions and models of assembly operations. From this layout, the simulation model was automatically generated and simulation was executed according to the input data. The aim was to verify the models and the simulation program, so that the measured output data and results of simulation could be compared. No significant deviation between the measured and the simulated results was found. It is therefore concluded that the LASIMCO program can be used for the planning of assembly lines.

The company's lines have asynchronous transport systems and for some cases their outputs can be calculated (using the queuing theory or Markov chains). If the model includes employees who remove defective parts or recover the stations, it becomes highly non-deterministic. Here the only way of obtaining the actual output is to simulate the dynamic behaviour of the line. Since the company uses AutoCAD (and a special library of drawings of standardised assembly modules) in drawing the layouts of their assembly lines, LASIMCO was accepted well and beneficially implemented as a handy integrated layout development tool.

5.3 Simulation in Robotics – ROMOBIL/SITAR

This chapter is dealing with a new approach in robot simulation – a PC-based, user-oriented simulation package for robotised assembly cells. The system includes two different packages, called ROMOBIL and SITAR. The interactive modelling module ROMOBIL allows the definition of the robot cell within CAD-environment. To meet the requirements of a 'low-cost' solution – e.g. reduced hardware costs – and because of being one of the most successful CAD systems at the moment, software package AutoCAD was selected as environment and interface for modelling module ROMOBIL. The second part of the system, the task evaluation module SITAR, includes the calculation of the cycle time of the assembly process combined with a 2D-animation of the movement of the robots Tool-Centre-Point. Like ROMOBIL, simulation module SITAR is embedded into CAD-software AutoCAD.

In addition to a short description of the developed simulation system, an application example of the package will be presented in this article. This study is dealing with the automated assembling of moisture-proof lamps. The assembly process includes the handling and processing of seven different parts as well as two screwing processes. Three cell-layouts have being developed and evaluated using simulation package ROMOBIL/SITAR.

5.3.1 Introduction

When planning a robotised (assembly) cell, one has to solve different, complex tasks. In most cases, development of the cell layout is still carried out by 'trial-and-error' method. Using *simulation technique* as a planning tool in the area of flexible assembly systems brings various advantages:

1. It can be used as a planning tool in robot design by reducing the need for physical prototypes and allowing simulated trials to be carried out. The time and cost spent for robot design can be reduced.

2. The simulation system can be utilised to choose a suitable robot for a given

assembly task. Robot selection can be facilitated by being able to visualise the robot geometry and work envelope, estimate cycle times, and consider various other parameters, such as joint movement limitations. Assuming the simulation package contains a library of different robot models, the user has the opportunity to make comparisons between them.

3. Robotised cells can be simulated and modified in order to achieve the most efficient work cell layout for each application. Potential collisions can be detected and robot arm movements optimised.

4. Simulation is an efficient method for education and training in robot-ics. This allows the user to program the robot and experiment with layouts without the danger inherent in using the 'physical' cell components.

Conventional methods for cell-layouting and programming usually lead to a massive set-up time for the cell when changing the type of product to be assembled. As a matter of fact, simulation is an important contribution for solving the problems mentioned above. With the technique of simulation it is possible to layout and program the assembly cell by means of a graphical model instead of using the real system. Concerning the problems in planning a robotised assembly cell, the basic tasks of a simulation software are:

- modelling of the cell and its components;
- programming the cell (and the robots);
- simulation/animation of the assemby process;
- sophisticated user-support with graphical man-machine interfaces.

Although much research is still being carried out in this area a number of commercial packages are available (GRASP, PLACE, CATIA-ROBOTICS, ROBOCAM, ROBOGRAPHICS, ROBCAD, etc.). Running on powerful (and costly) workstations and graphic systems, these simulation systems are offering wide-spread features. Using simulation only to check-up the assembly process and to calculate (better: to estimate) its cycle time, these full-blown simulation packages are sometimes rather too complex. Furthermore, these programs are very expensive (soft- and hardware) – especially small and medium sized companies are not able to afford those kind of simulation studies.

Another way is to use a (commercial) general purpose simulation system (e.g. SIMPLE++, SIMAN, GPSS or others). An advantage of this solution is the possible combination of the robotised cell with its environment (transport systems, shop control system, etc.). The missing support in detailed modelling of robotics features with general purpose simulation packages is one of the significant drawbacks of this solution [G.90].

To combine the pro's of the solutions mentioned above, the Institute for Handling Devices and Robotics at the Technical University of Vienna has been working on a 'low-cost' simulation system for robotised assembly cells – ROMOBIL/SITAR. This package has being designed as an AutoCAD application module. Program handling and user interfaces are following the 'standards' set by this widely used CAD package

– the full set of AutoCAD drawing features can be used by ROMOBIL/SITAR. Choosing AutoCAD as a 'carrier' as well as focusing to a simulation of *assembly steps* (instead of investigating robot's *movement* per se) also allows the use of PC's as the system hardware.

5.3.2 Simulation System SITAR

Beside the particular requirements for an inexpensive simulation system for utilisation in small (and medium) sized companies, following aspects were taken into account when planning this simulation package.

Open system

Generally, simulation systems (for robotics) are self-contained. An interface to subprograms for planning robotised assembly cells – e.g. to a software package for an efficient, interactive selection of components – will be supported not very often. The simulation system SITAR enables the utilisation of planning data from external packages. For example, an interface to a cell-planning system called ROBPLAN (developed at the Institute for Handling Devices and Robotics at the Technical University of Vienna [KPG95]) has been realised.

Modelling and simulation/animation takes place within a CAD-software package – the geometric data of the cell layout can be used directly. The combination between cell-layouting (CAD-functions), modelling (definition of the assembly sequence), and simulation (optimisation of the assembly sequence) enables an interactive planning and optimisation within an integrated environment (Fig. 5.7). The software package

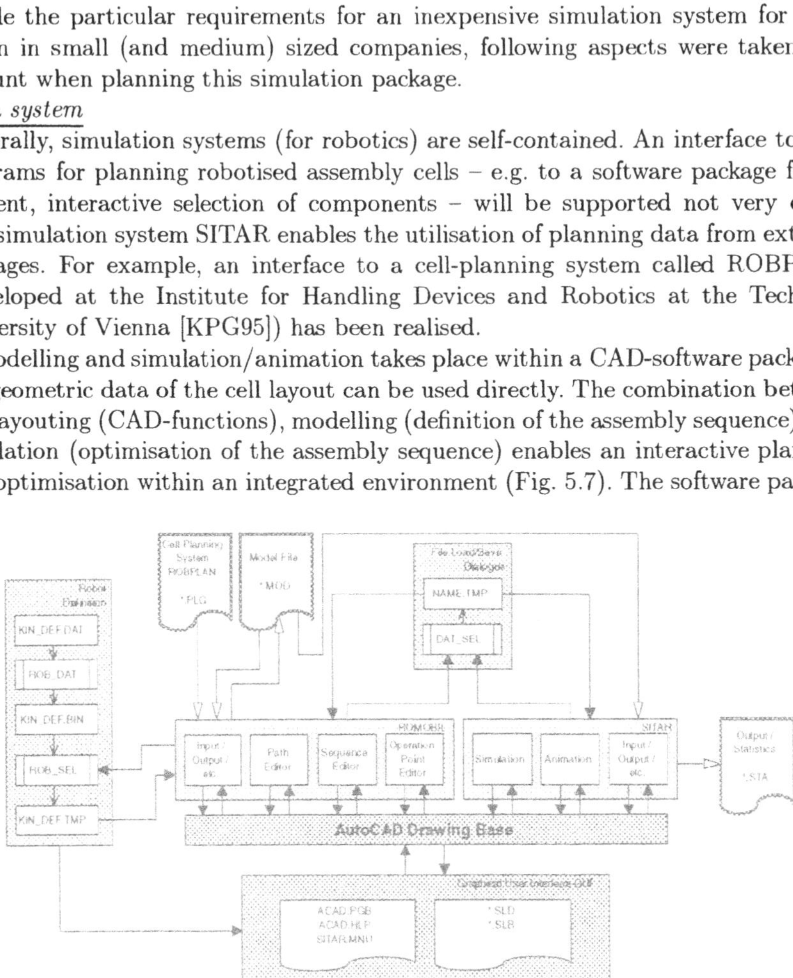

FIGURE 5.7. System structure of the simulation system ROMOBIL/SITAR.

consists of different modules – each of them servicing the AutoCAD drawing base. This results in a construction **without** interface between the modules itself – but with well-defined interfaces between the particular module and the drawing entities (processing their DXF-codes). If there is a need for a new functionality, new modules can be defined in that way and grouped together into a new application. With this, system ROBPLAN can be extended **without having** to recompile the standard source.

User-oriented interface

When designing the simulation package ROMOBIL/SITAR, the developers have tried to realise a new approach in modelling of robotised processes. Here, the planned **assembly sequence** – consisting of robot movements, tooling or handling operations, communication with and between peripheral devices, etc. – is the centre of the simulation model. The goal of the simulation is not to analyse the path (or the trajectory) per se. Furthermore, simulation should constitute an aid for calculation and evaluation of the paths and their optimal sequence as part of the entire assembly process. Since simulation in robotics can be seen as a combination between *discrete event* (e.g. starting of particular robot movements) and *continuous* (robot movement per se) simulation, system ROMOBIL/SITAR is focused more on the discrete event point of view.

Modelling of the assembly schedule, robot trajectories, and working positions is realised in a graphical manner. The required model entity – either being 'discrete' type (start/endpoint of a trajectory, communication signal between cell components, etc.) or 'continuous' type (trajectory) – has to be selected and – after definition of the specific parameters of this entity – inserted in the layout ('Pick-and-Place' method).

The simulation system ROMOBIL/SITAR includes two different software parts:

- **ROMOBIL (Interactive Robot Cell Modelling System)**
  ROMOBIL enables a user-oriented, interactive modelling of the robotised cell within CAD-environment. Furthermore, this modelling system allows a perfect connection between simulation module SITAR and a planning software for robotised assembly cells 'ROBPLAN' [KPG93]. Because of its numerous installations and the demands for a 'low-cost' solution (concerning hardware as well as software), CAD system AutoCAD was selected as an environment and interface for our simulation system.

- **SITAR (Simulation System for Evaluation of Robot Cells)**
  This software package serves for a rough calculation of the cycle time of robotised assembly processes (modelled with ROMOBIL or other programs) combined with a 2D-animation of the Tool-Centre-Point of the robot(s). Beside animation of the motion of the robot effector, several data – e.g. duration of each assembly step, utilisation of the robots, delay time, etc. – are displayed in order to support the user. Like ROMOBIL, the simulation module SITAR is running with CAD-software AutoCAD.

5.3.3 Modelling of Robot Cells in ROMOBIL

Geometric modell

One of the basics of simulation studies is an existing geometrical model of the robot cell. In our simulation system ROMOBIL/SITAR the geometrical model – i.e. the cell layout – can be created using standard AutoCAD drawing functions. If there is a connection to ROBPLAN, the phase of cell-layouting is reduced to the arrangement of the cell components (selection and drawing of all required components is part of the ROBPLAN functions) [KPG93].

Kinematics modell

The kinematics model usually contains the definition of the robot, e.g. number of DOF, type of each robot link, work-space, maximal velocity and maximal acceleration of each link. For solving the direct and inverse kinematics of manipulators it is necessary to describe both, the type of the link and its connection to a neighbouring link. The definition of mechanisms by means of these quantities usually follows a convention called the 'Denavit-Hartenberg Notation'. SITAR should be able to simulate each (commercial) available industrial robot in (at least) 'real-time'. Therefore – and because of the moderate requirements regarding the accuracy of the calculated cycle time – it was determined to simulate just the gripper guide gear (the first three DOF). Starting with modelling the user has to select a robot structure from six possible (and industrial mostly used) kinematics structures (TTT, RTT, TRT, RRT, RRT-Scara, RRR; 'R' ... rotational link, 'T' ... translational link) and to define the link lengths and eccentricities. After this definition of the kinematics model the robot(s) can be placed in the work cell and the user can start to define the manufacturing/assembly program.

Programming

Creation of the program takes place with an interactive dialogue with a graphical, menu- and window driven interface (Fig. 5.8). One has to select the required model entity (start/end point, path, sequence), scale it by setting the entity specific parameters, and arrange it in the cell layout (pick-and-place method). For modelling of the robot paths, the user can choose between (co-ordinated) point-to-point motion (PTP), continuous path motion (CP) and circular path (CIR). Beside the definition of path type and relative speed of the path (monitor speed), it is possible to specify a delay at the end-point of the path (related to the required time for operation in this point). To simulate the co-ordination between different robots and/or robot and peripheral devices, a signal interchange can be modelled as well. Every inserted model entity will be recorded in a readable and editable file. This file serves as security backup for the model, for better documentation of the scenario and as an input file for the cell simulator module SITAR.

Modelling using the data from cell-planning with ROBPLAN

One of the basic features of the simulation package ROMOBIL/SITAR is the interface to the robot cell planning software ROBPLAN. As a result of the planning process, ROBPLAN generates a data file which contains the kinematics definition of the used robots and a detailed description of the single assembly steps and the used devices (gripper, tool, magazine). During modelling process, these data were used as improved support to the user.

- Positioning of the 'operation-points': First step is to read all operat-ion-points from data file and insert them (ordered) into a table. During modelling phase, the user is prompted to insert them into the cell-layout (i.e. define their co-ordinates in Cartesian work space).

- Definition of the robot paths: In second phase of modelling, each assembly step is read from the input file. With this information a suitable prompt will be created in order to support the user for the definition of the trajectories.

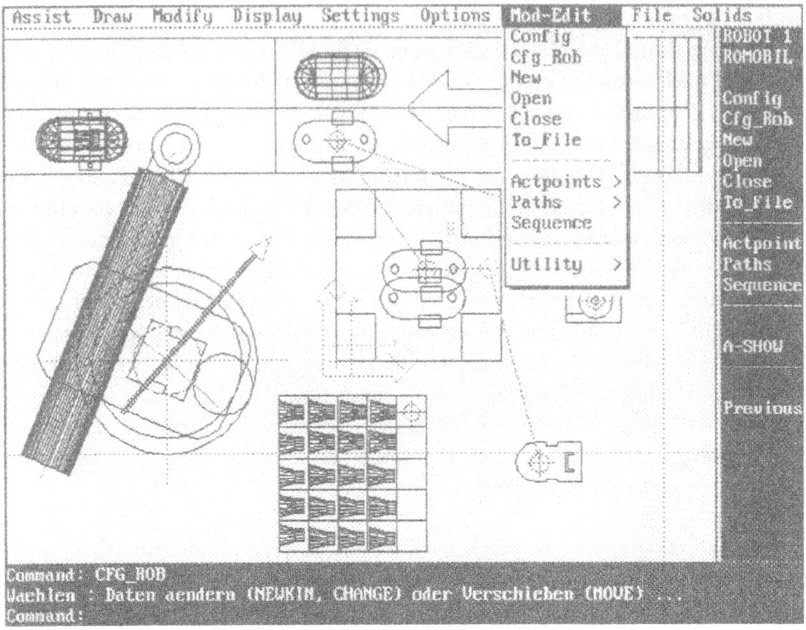

FIGURE 5.8. Cell layout including simulation modell.

5.3.4 Application example

To check the qualification for industrial applications, simulation system ROMOBIL/SITAR has been utilised for planning an automated assembly cell for moisture-proof lamps. The assembly process includes handling and processing of seven different parts as well as two screwing processes. Three cell-layouts have being developed and evaluated in order to represent the robotised cell and the assembly process as well as to optimise the arrangement of the cell components and the assembly sequence (Fig. 5.8).

<u>Results</u>

Beside execution and evaluation of different simulation scenarios in order to increase the efficiency of the assembly cell, this application example was planned to show the limits of the developed simulation system. Resulting from the experiences and problems during modelling and experimental phase the following conclusions can be drawn in detail:

- Regardless the limited amount of memory (PC-platform!), the presented simulation system allows the analysis of *medium-scaled systems* (i.e. about 100 assembly steps) in a very comprehensible and efficient manner.

- Checking of possible *workspace violations* during arrangement of the cell components (during modelling phase!) turns out to be a very efficient tool for layout planning.

- The implemented signal interchange between two *co-operating robots* enables an improved tuning of the assembly sequence.

- For modelling and simulation of even more complex systems, the graphic-oriented user interface of the modelling system lose some of its clearness. To solve this problems additional *alphanumeric modelling features* as well as *improved zoom functions* are planned for the next release of ROMOBIL/SITAR.

5.4 Conclusion

Quality of planning extremely affects the overall costs of a manufacturing system. Considering the decreasing life-time of certain products, saving of time in the planning stage will become more and more important. Small and medium sized companies are only able to take an advantage from their possibilities to react to changing market conditions, if also their production equipment provides with flexibility. To help with these problems, there is a method at hand the significant benefit of which has already been proven: **simulation.**

Especially for simulation of robotised manufacturing systems, there are several tools available, specialised on 'high-tech simulation' – but too expensive regarding to hard- **and** software costs (ROBCAD, IGRIP, CATIA-ROBOTICS). For simulation of discrete event systems there are systems on the market, which are allowing simulation on a more abstract level, but on the other hand not beeing able to meet the special requirements of robotics.

At the Institute for Handling Devices and Robotics (IHRT) at the Technical University of Vienna, a PC-based 'low-cost' simulation system for robotised assembly cells **ROMOBIL/SITAR** – has being developed. The system allows a very fast – but also sufficiently precise – evaluation of the assembly process. The features of the system are:

- interactive modelling, simulation and animation of the robot(s) with-in CAD environment
- interface to robotics programming languages VAL, BAPS, and AML (in preparation)
- import of robot data with IRDATA (in preparation)
- interface to assembly cell planning software **ROBPLAN** (developed at IHRT)
- simulation of the co-operation of two robots
- open, user-oriented, menu-driven simulation system
- advanced modelling concept ('Pick-and-Place')

Since ROMOBIL/SITAR has been designed as a widely open software package, it enables further extensions. For instance some new features according to improved modelling, task oriented programming, automatic path planning (better: sequence scheduling), interface to the system for 'Hierarchical Control of Assembly Cells **C_CTRL'** (developed at IHRT) are in preparation.

LASIMCO – a simulation system for planning of assembly systems developed by the Laboratory for Handling and Assembly Systems Automation, University of Ljubljana – is another example for a 'low-cost' solution, which successfully integrate CAD and simulation technique into one system.

The program LASIMCO has been used for real problems from the industry as well as for practical lectures in the field of handling and assembly systems. Its usage shows, that it is a powerful tool for analysis of any discrete event driven systems, because of direct use of system model in simulation and because of suitability of Petri nets theory for modelling of systems with concurrency and mutual exclusion. Further work will be focused on implementation of layout and simulation model generation for other types of assembly systems and on development of a modelling tool for the user to be able to generate his own objects and Pert net models of the object in a graphical way. Further extensions are also planned in order to increase simulation speed and to enable 3D animation technique as well as to extend the conception of simulation with the idea of virtual manufacturing systems [TD96].

5.5 References

[C.92] Lueth T. C. Automated computer-aided layout planning for robot work-cells. In *Preprints of the IFAC Symposium on Information Control Problems in Manufacturing INCOM'92, Toronto, Canada, Volume II*, pages 672–677, 1992.

[G.90] Kronreif G. Contribution to simulation technique in the area of automated assembling, 1990.

[G.94] Kronreif G. *Interactive Simulation System for Robotized Assembly Cells.* PhD thesis, Inst. for Handling Devices and Robotics, TU Vienna, 1994.

[J.L81] Peterson J.L. *Petri Net Theory and Modeling of the Systems.* Prentice-Hall, Inc., 1981.

[KPG93] Kratschmann R. Kopacek P. and Kronreif G. Computer aided design of robot-equipped assembly cells. *e&i*, 110:359, 1993.

[KPG95] Noe D. Kopacek P. and Kronreif G. Semiautomatic knowledge based planning of small assembly cells. In *Proceedings of the 1st World Congress on Intelligent Manufacturing Processes and Systems IMP&S'95, Mayaguez, Puerto Rico, Vol. 1*, pages 574–582, 1995.

[RA90] Al-Jaar R.Y. and Desrochers A.A. Performance evaluation of automated manufacturing systems using generalized stochastic petri nets. *IEEE Trans. Robotics and Automation*, 6(6):621–639, 1990.

[SS94] Valavanis K.P. Ramaswamy S. and Landry S.P. Extended petri net-based modeling, analysis and simulation of intelligent materials handling system. *Journal of Intelligent and Robotic Systems*, 10(1):79–108, 1994.

[T.89] Murata T. Petri nets: Properties, analysis and applications. *Proceedings of the IEEE*, 77(4):541–580, 1989.

[T.93] Perme T. The communication model of integrated manufacturing systems. Master's thesis, Faculty of Mechanical Engineering, University of Ljubljana, 1993.

[TD94] Perme T. and Noe D. Simulation model of a pallet assembly system. In *Proceedings of the 3rd International Workshop on Robotics in Alpe-Adria RAA'94, Bled, Slovenija,*, pages 75–80, 1994.

[TD96] Perme T. and Noe D. A low-cost solution of virtual manufacturing systems. In *Preprints of 13th IFAC World Congress 1996, San Francisco, USA, Vol. B*, pages 299–304, 1996.

[YP93] Van Brussel H. Peng Y. and Valckenaers P. Modelling flexible manufacturing systems based on petri nets. *Annals of the CIRP*, 42(1):479–484, 1993.

6

Model of Material Handling and Robotics

Chin-Yin Huang[1]
Shimon Y. Nof[1]

6.1 Introduction

In today's manufacturing systems, robots and material handling facilities flexibly circulate and integrate material flows among distributed, automatic production equipment. Robots and material handling facilities provide two important and unique capabilities into manufacturing systems: (1) dynamic and mixed material flows, and (2) flexible relationships between manufacturing systems and their environment. The first capability is based on the reprogrammability and intelligence of modern material handling devices and robots. Hence, the material flows are changeable rather than fixed. In parallel, because customers, competitors, and suppliers expect the availability of such an integrated and flexible manufacturing system, the second capability is developed.

The study of models of material handling and robotics (MHR) has been divided into two levels: micro and macro [17]. Research in micro level studies the economics of robotic operations, whereas research of robotic installation and configuration is categorized in the macro level. At the "micro" level, the individual operational characteristics of robot(s), e.g., cycle time, payload, with respect to various tasks are studied. On the other hand, at the "macro" level, the individual robots are not analyzed, relationships between robots and other facilities, e.g., machines, AGVs, aisles, etc. become the research focus. Mathematical programming, heuristic algorithms, queuing networks, and simulation are the most common traditional techniques that have been applied in modeling MHR systems. In addition, the objective of the models is usually to minimize costs or distance of material flows [15]. These approaches are classified as "traditional approaches", yet, they are not considered to be inferior to the "non-traditional" approaches described below.

Modern computer and communication technologies can improve the quality, accuracy, and timeliness of decisions obtained with the traditional approaches, and can also provide better design tools for MHR. In this chapter, a tool perspective is applied to study models of MHR. Two integrated tools for MHR are introduced: *facility description language* (*FDL*) and *concurrent flexible specification* (*CFS*). The

[1]School of Industrial Engineering, 1287 Grissom Hall, Purdue University, West Lafayette, IN 47907-1287, USA, TEL: (765) 494-0289, FAX: (765) 494-1299, e-mail: nof@ecn.purdue.edu

two tools represent the newer perspectives which are not considered in traditional modeling approaches.

This chapter is organized as follows. The traditional modeling approaches are first reviewed in section 2. Based on the evolution of traditional approaches, several concerns are discussed as background to the necessity of using the tool perspective to study the models of MHR. Certain models of MHR in tool perspective are also introduced and compared with FDL and CFS in this section. Then, FDL is described in section 3, and CFS is introduced in section 4. Finally, discussion and conclusions are presented in section 5.

6.2 Traditional Models of Material Handling and Robotics

6.2.1 Traditional Approaches

As mentioned in section 1, the economics of robotic operations are the main issue in the micro level modeling of MHR. Typical examples are the Robotic Time and Motion (RTM) technique [11] and part flow in the robotic assembly plan problem [12]. Basically, the goal of such techniques is to provide operational economic models and performance measures of the robotic operations, so the tasks can be effectively and efficiently analyzed, improved or optimized, and then performed. The focus of the models are *robots*, *material handling facilities*, and *tasks*.

The modeling approaches in micro level usually provide a general framework, simulation or mathematical models, and principles based on the accumulation of statistical data from the operations of MHR. Then, when a new case occurs, the principles are applied to predict a suitable decision about the flow operations.

On the other hand, the macro level modeling studies the economics of larger-scale robotic installation and configuration. In the macro level, production and flow layouts are also considered in the system design. Modeling at this level includes equipment selection, AGV path design, palette size and load analysis, capacity analysis of material handling [2] [15], work transportation [8], and location analysis [17].

Mathematical programming, heuristic algorithms, queuing networks, and simulation are also the main techniques that have been applied in modeling and solving problems in the macro level. By selecting the variables for the problems appropriately, a mathematical model (e.g., linear programming model) with objective function(s) and constraints is constructed. Then, an optimal solution is searched and obtained. Heuristic algorithms play a similar role to a mathematical program. However, heuristic algorithms eliminate the efforts in searching for the solutions.

In queuing network models of MHR, the throughput and utilization of the whole system are usually the main objective. The capacities of MHR and machines, which are related with the service rate, become the decision variables. The functions of simulation and animation are similar to the function of the queuing models, but simulation and animation provide a more flexible and visible modeling environment. The design of well-organized material flows to meet the customer requirements is the modeling objective.

6.2.2 Concerns with Traditional Modeling Approaches

When traditional approaches are used in modeling manufacturing systems, it can be observed that several problems are not considered:

1. *Conflicts among designers*: Large-scale system design usually is performed by people with different background knowledge. Each designer is typically responsible for only part of the system (subsystem). With traditional modeling approaches, it is impossible, or very difficult to identify and consider the conflicts that may exist among the subsystems.

2. *Constraints in physical environment*: the abstracted variables in the traditional modeling approaches cannot totally reflect the material flows in the real world. The implementation of the decisions may create problems because of the constraints in physical environment, e.g., collision, passages that are too narrow, etc. Thus, many engineering changes, delay, and financial loss can result.

3. *The information flows in the manufacturing systems*: Traditional modeling approaches usually consider physical material flows and facility design. However, beyond the physical facilities, there are information flows which control the operations of the facilities. Well-organized information flows can control the material flows in an effective, efficient, and flexible manner. Therefore, it is necessary to also include in models of MHR the information flow perspective.

6.2.3 A Comparison of Models of Material Handling and Robotics with the Tool Perspective

Some researchers have noticed the importance of modeling MHR with the tool perspective. In this section, a sample survey is presented in Table 1. The three modeling concerns in the previous section are part of the criteria in the survey table. It is found that the concerns are not addressed by the first four models. The approach of the first four models still follows the traditional approaches, except for the second model.

6.3 Facility Description Language (FDL)

FDL provides an integrated modeling tool for distributed engineers working on various aspects of manufacturing system design [19][20][21]. FDL is implemented in a 3-D emulation environment–ROBCAD [18], which provides a dynamic and physically visible (comparing with "iconically visible" in simulation animation) CAD environment. Therefore, all the materials, operations of robots and material handling facilities, and machines are shown by their true physical relationships.

In the framework of FDL, the manufacturing systems are modeled by computer information and graphics. Various information items are included in the model, which is not seen in traditional models [19]:

1. Organizational relationship among facility components.

2. Specification of working locations for robots.

No.	Modeling approach/tool, reference	Methodologies Integrated	Measured criteria or goal of the model	Three concerns addressed?
1.	[9] Nadoli and Rangaswami	expert system, simulator	waiting time, congestion level, etc.	No
2.	[4] Chadha et al.	IDEF0, Data Flow Diagram, Extened Entity Relationship	develop an integrated information system	(3) only
3.	[13] Prakash and Chen	SIMAN IV simulator	speed of AGV and dispatching rules	No
4.	[16] Sly	AutoCAD with FactoryFLOW	Min. distance of material flows	No
5.	*FDL* [19][20]	ROBCAD and simulator	Integrated tool for distributed engineers	(1), (2), and (3)
6.	*CFS* [6][7]	Data/Control Flow Diagram, Petri Net	Integrated tool to incorporate material flows, information flows, and control signals	(1) and (3) only

TABLE 6.1. A Sample Survey of Models of Material Handling and Robotics with the Tool Perspective

3. Flow of parts and materials through the facility.
4. Location of personnel and equipment aisles.
5. Control relationships between devices.
6. Designation of sensors and their targets.

These information items are supported by the modeling functions of FDL (Table 2). An example of modeling an aisle by an FDL function is:

<u>*Aisle*</u> *Re – define aspects pertaining to an aisle.*
syntax ***aisle*** *action path parent size*

action char (A, C, D) *add, change, delete record (mandatory)*
path char[16] *path name (ROBCAD name)*
parent char[16] *parent name from device table*
size char "HUMAN", "AGV", "SMALL_FL", or "LARGE_FL"

The model of MHR in FDL becomes a list of syntax. The list of syntax triggers the ROBCAD program to construct a 3-D emulation model (see the model inside the window of Figure 1). Figure 1 is an example of FDL in ROBCAD system. The upper left window is used to input the information of the system including the geometric information of facilities, material flow information, and material flow control information. The lower left window is used to show the output information (e.g., a collision occurring on the robot during the material handling). In addition, FDL provides a reconciliation function (see the right function menu in Figure 1). Therefore, all the conflicts on the material flows can be resolved according to the build-in algorithm. The reconciliation function may change the positions of robots or machines to avoid the collision or unreachability of material handling.

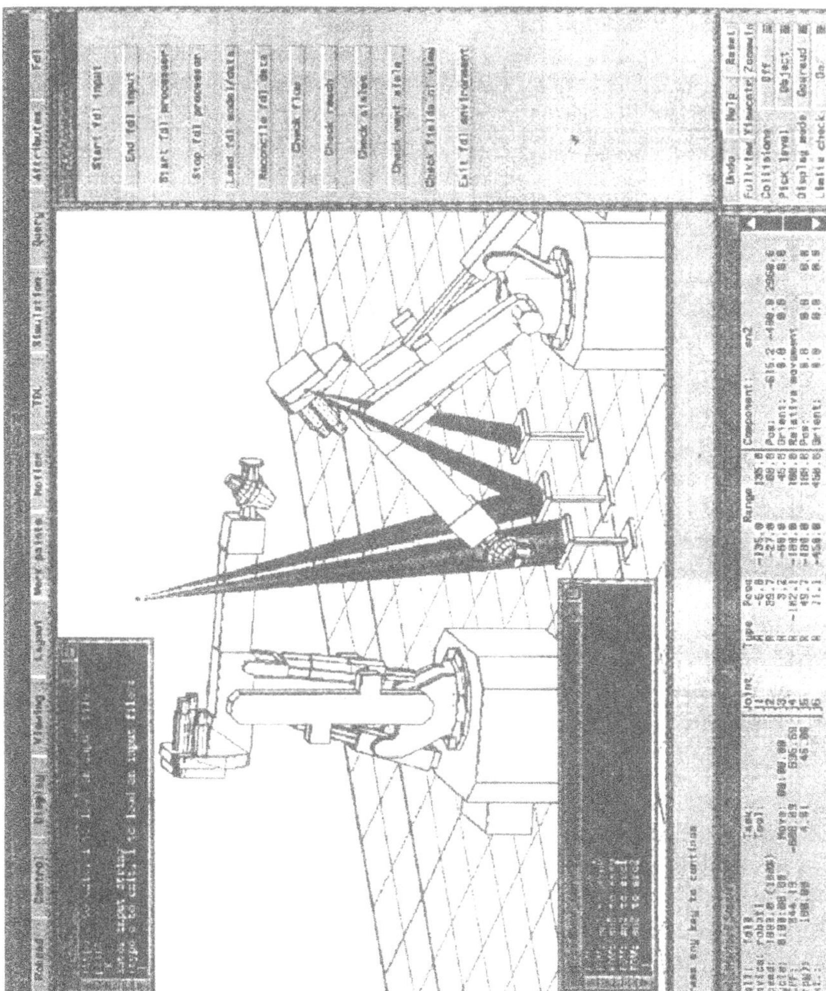

FIGURE 6.1. An Example of Facility Description Language in ROBCAD System

Function Class	*Function*
FDL Definition Input Functions	Aisle, Capability, Control, Define (aisle, path, perimeter), Device, Facility, Part, Process, ProcessPart, Sensor, Transfer, Workpoint
FDL Manipulation Input Functions	Attach, Delete, Detach, Display (top, lower, bottom, name), Moveback, ShiftBy, ShiftTo
FDL Utility Input Functions	Comment, Print, Save
FDL Evaluation Function	Reconcile Database, Evaluate Aisles, Evaluate Device Reach, Display Material Flow Paths, Evaluate Fields of View

TABLE 6.2. FDL Modeling Functions

Since FDL provides such direct syntax specifications, the designers can use the syntax to model and develop their subsystems. When the designers are in different locations, their subsystems can submit input to the host ROBCAD system to construct the entire system, then use the reconciliation function to adjust the subsystems if conflicts occur. Therefore, the cooperation of designers in different locations for different subsystems can be achieved in FDL.

6.4 Concurrent Flexible Specifications (CFS) for Material Handling and Robotics

By specification, engineers describe the way that a system should be constructed. Concurrent, flexible specifications for MHR, in other words, are provided by several engineers to model an MHR with flexibility to design changes. In a manufacturing system, there is a physical flow of raw materials, parts and subassemblies, together with an information and control flow consisting of status (system state) and control signals. The control and status signals govern the behavior of the physical flows [10]. In order to simultaneously achieve optimal capacity loading with maintained or increased flexibility, an exact definition of material and information flow becomes necessary. This approach is followed by the CFS modeling [6][7]. The specification should not only represent the logical structures of different functions, but also their logical connections, e.g., the structures of material and information flow in a manufacturing cell [5].

Another important requirement of specification is the ability to precisely define the real-time behavior of the system. In a real-time system many of the inputs to the system are signals that indicate the occurrence of events. These inputs do not pass data to the system to be processed. Generally they occur in streams over time and their purpose is to trigger some process in the system repeatedly. Furthermore, many systems are made up of subsystems, any of which may be active or non-active at a particular time during system operation. For this reason, the treatment of timing is an essential element of the specification.

To integrate the above requirements, tools which can incorporate material flows, information flows, and control signals are required. Therefore, different representations of specifications through different tools should not be independent or mutually exclusive, but should support each other by forming a concurrent, comprehensive specification of the system. For the model of MHR, the specification of *functional* and *real-time logic* is of importance since these attributes describe what and how the system is executing. This information is necessary to determine "how" the processes are to be implemented with physical equipment. By utilizing two complementary representations, both these aspects of system behavior can be specified concurrently.

6.4.1 CFS Using Data/Control Flow Diagram and Petri Nets

Data/Control flow diagrams (DFD/CFDs) which are enhanced with real-time extensions when used in conjunction with Petri Nets, provide a suitable framework for concurrent specification of functional and real-time state logic. The main reason supporting this fact is the ability to maintain identical structural decompositions in both representations at all levels of detail in the specification. This model is accomplished by maintaining identical partitioning of processes in both specifications.

With DFD/CFDs, partitioning is accomplished by hierarchical decomposition of bubbles that represent processes or tasks. An identical hierarchical partitioning can be created with Petri Nets by representing processes with sub-nets at higher levels and then showing the detailed, internal net at lower levels of detail (demonstration is shown in section 5). The DFD/CFDs provide a process definition model of the system while the Petri Nets provide a process analysis model for the study of real-time state behavior. Even though object oriented modeling becomes a more popular technique of system design, Data/control flow diagram is still an acceptable technique in our case study. Researchers have proved the possibility to transform data flow diagrams to object models (e.g., [1]). However, both transformation procedures and object oriented modeling are beyond the discussion scope of this chapter.

Both these techniques are realized by two software packages: Teamwork and P-NUT. Teamwork is a CASE (Computer Aided Software Engineering) tool family that automates standard structured methodologies using interactive computer graphics and multi-user workstation power. P-NUT is a set of tools that are developed by the Distributed Systems Project in the Information and Computer Science Department of the University of California at Irvine [14] to assist engineers in applying various Petri Net based analysis methods. Both software packages are described in the following section.

6.4.2 Overview of Specification Software Tools

- Teamwork:
  Teamwork is an integrated tool set to build, store, maintain, and review structured specifications for functional system analysis [3]. Two of the important analysis tools in Teamwork are illustrated:

  1. Teamwork/SA is an environment for system analysis. It enables analysts to rapidly create and verify functional system specifications.
  2. Teamwork/RT is an environment for real-time modeling. It is an extension of Teamwork/SA and allows analysts to model the complexities of real-time systems, including real-time sequencing, timing and control.

  For Teamwork/SA and Teamwork/RT, there is a graphical DFD/CFD editor for creating *bubbles*, *stores*, *terminators*, and *data flows*, and a P-Spec (Process Specification) editor for supporting the creation of textual process specifications. Control flows in the DFD/CFDs are distinguished from data flows by being represented as dashed lines. When performing real-time analysis, C-Specs (Control Specifications) specify the processing of control flows as events or actions through state transition diagrams, state event matrix, process activation tables, or decision tables.

- P-NUT:
  There are three objects created in P-NUT: *Petri Nets*, *Reachability Graphs*, and *Execution Traces*. Petri Nets are accepted by the system in textual form (circle 0 in Figure 2) and transformed into an external representation (circle 1 in Figure 2). It is this internal representation that drives all remaining tools.

  Reachability graphs (circle 2 in Figure 2) are representation of all of (or part of) the state space of a Petri Net. There are graphs whose nodes represent states and whose edges represent state transitions. Execution traces (circle 3 in Figure 2) are representations of portions of the state space. An execution trace is one long path through the state space, where the same state may be visited many times. Each execution trace includes the complete specification of the Petri Net.

6.4.3 Case Study Application

In this section, a case study of applying CFS through the tools of Teamwork and P-NUT in an automated (robotic) assembly line for the production of alternators is presented. The production facility is divided into two separate assembly lines: (1) assembly line for stator-regulator end (SRE) and (2) assembly line for drive end (DE). Both lines are independently controlled and SRE subassemblies are transferred onto the DE line, where the final assembly, quick check, and final pack-out occur. The DE line is chosen for the illustration of CFS for the interesting control problems presented by the transfer station, and the presence of final assembly operations and pack-out.

Figure 3 shows a logical layout of the DE line, with the stations numbered. The arrows indicate the flow of pallets along the line. All operations with the exception of

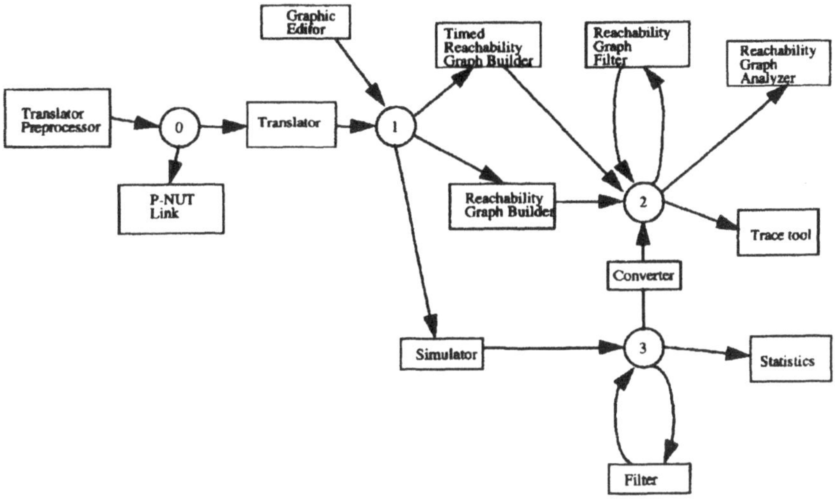

FIGURE 6.2. The P-NUT Suite of Tools [14]

quality check and pack-out are automated and controlled by a single main controller. The stations are fed with parts by bowl feeders and off-line pallets. At each station, parts are retrieved from bowl feeders (sensed optically) and the robot or other piece of fixed equipment waits for a pallet to arrive on the conveyor before performing their tasks. When the pallet arrives at the station, it is sensed by the magnetic sensor and the part on the pallet is sensed by an optical sensor. The optical sensor sends a signal to the main controller, which in turn sends a signal to the station, signaling it to execute its task. After the task is completed, the station responds to the main controller with a "DONE" signal.

Concurrent Specification of Assembly Line

A hierarchical specification is constructed with DFD/CFDs, and a detailed specification of the transfer process is represented through Petri Net and DFD/CFDs. The detailed specification illustrates the ability to produce a concurrent, flexible specification of process and (sub) system logic.

- Top-most level:
  The top-most level of the DFD/CFD specification is shown in Figure 4. This level is known as the Context Diagram. It represents the entire assembly process by a single bubble, and the connections to external entities. The dashed lines (control flows) represent control flow, while the solid lines represent physical objects that are inputs or outputs of the system.

- Level 1:
  The next level of detail models the entire process as being represented by four separated sub-processes (Figure 5). It should be noted that the identification

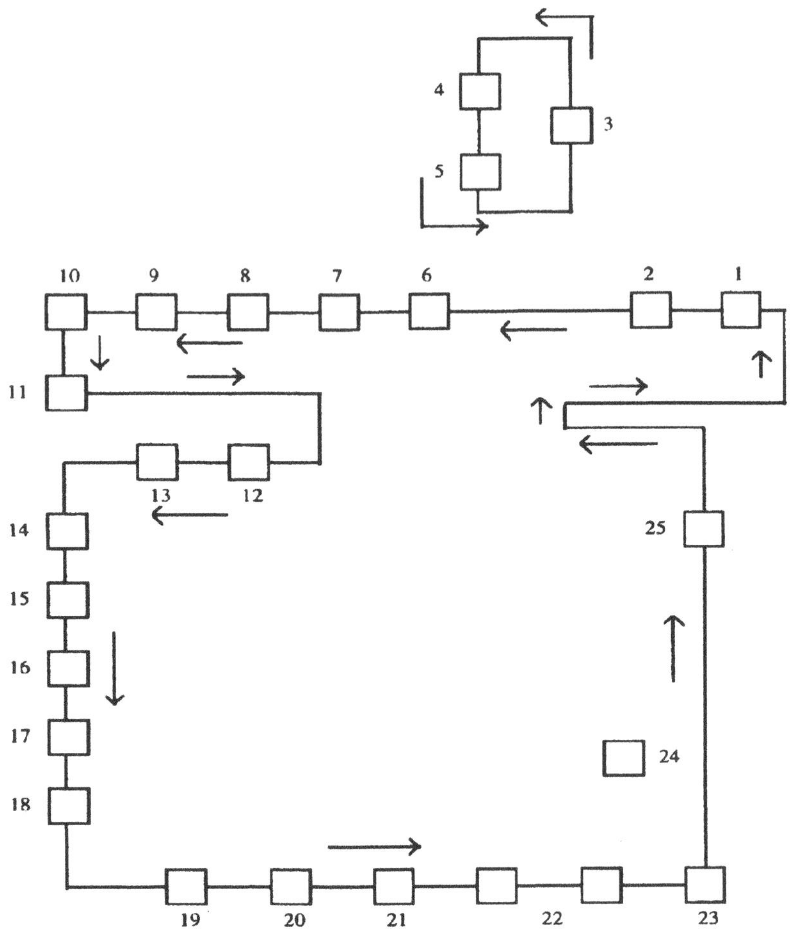

1. Place ROTOR
2. Place COLLAR
3. Load DE casting onto PALLETS
4. Press BEARING into DE casting
5. Spin BEARINGS
6. Place DE casting from OFF-LINE
7. Place FAN COLLAR, PULLEY
8. Rotate
9. Place NUT
10. Torque NUT
11. Transfer SRE
12. Place DE on SRE
13. Seat DE
14. Turn over
15. 3 THRU BOLTS
16. Torque BOLTS
17. Set BEARING
18. Rotate
19. Place COVER
20. Drive LOCK PINS
21. Stamp BC
22. TWO STAMP DATE
23. QC CHECK 1
24. QC CHECK 2
25. PACK-OUT

FIGURE 6.3. Logical Layout of the Drive End (DE) Robotic Assembly Line

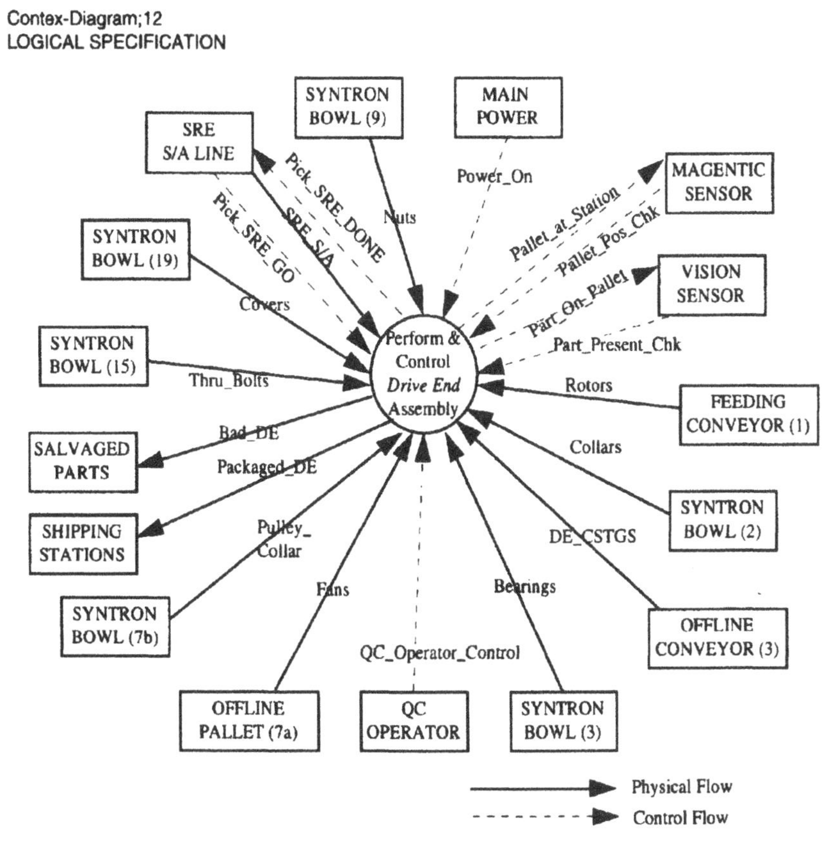

FIGURE 6.4. Context Diagram – Teamwork Assembly Line Specification

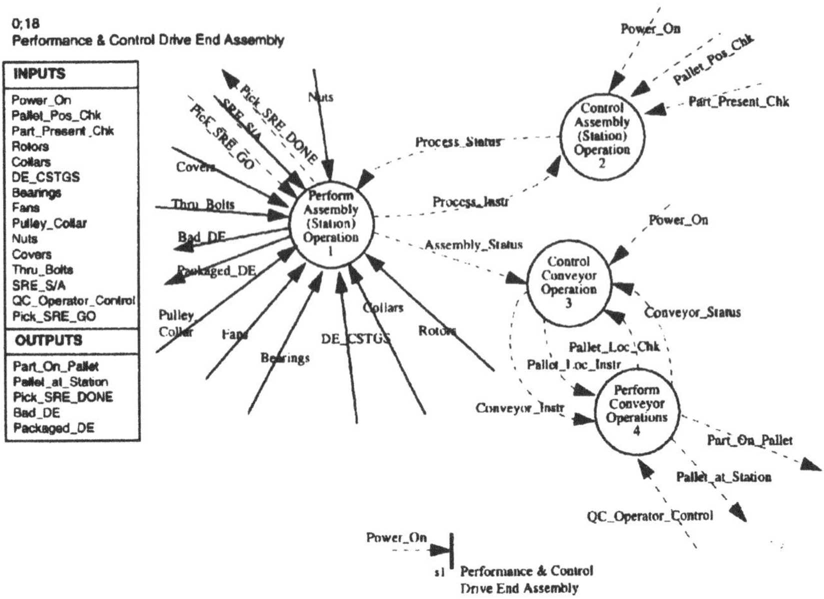

FIGURE 6.5. Level 1: Performance & Control Drive End Assembly

numbers in Figure 5 do not imply any prioritization or ordinal execution of the processes. The numbers are used only as a means of reference for subsequent levels of detail. Additionally, number "s1" is related with the *Process Activation Table* which will be explained later.

- Level 2:
  Figure 6 shows part of the examination of the process "Perform Assembly (Station) Operation" at Level 2. Transfer SRE is the main activity. Detailed contents of Transfer SRE is illustrated in Level 3 (Figure 7).

- Level 3: Detailed "Transfer SRE" is depicted in Figure 7 by displaying the child diagram for the process. The robot at the transfer station receives two signals: one from the SRE line (Pick_SRE_GO) and one from the DE line (PLC_SRE_GO). The transfer process is controlled so that when the robot is idle, it is always at its home position with an SRE subassembly in its grasp, and waiting for a DE pallet to arrive. There is also a transfer buffer where SRE subassembly are placed and picked. Depending on the type of signal or combination of signals received by the robot, there are three different actions which it can perform. These actions are illustrated by the bubbles 1, 2, and 3 in Figure 7.

  The C-Spec at the bottom of Figure 7 (vertical line) identifies the control flows that determine the activation of the processes. A process activation table (PAT) that describes this control logic is shown in Table 3. PAT is displayed by

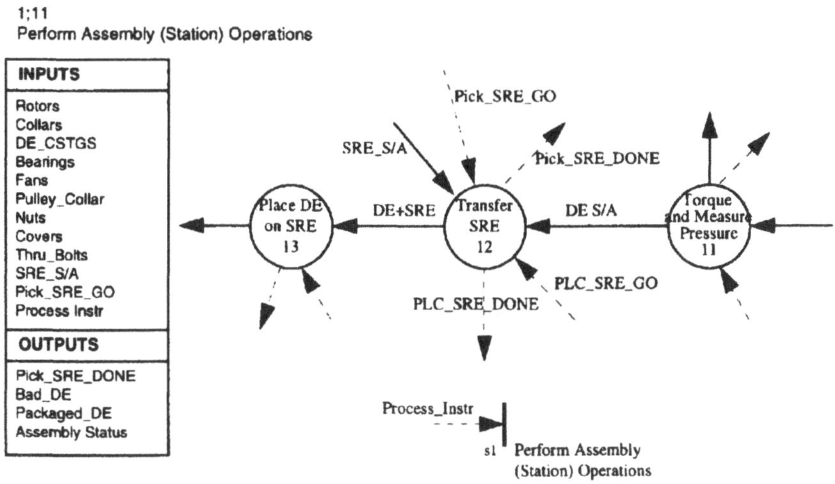

FIGURE 6.6. Partial DFD/CFD of Perform Assembly (Station) Operation in Level 2

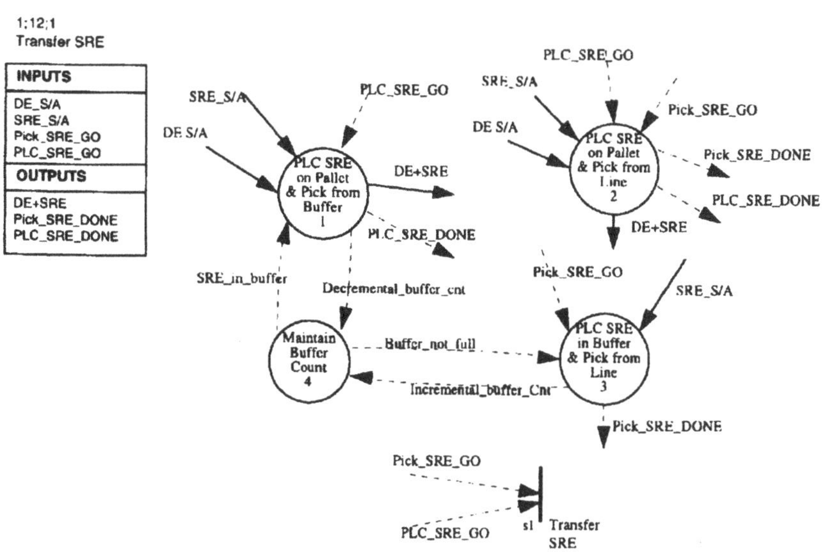

FIGURE 6.7. Level 3: Transfer SRE DFD/CFD

1.12;s1;1
Transfer SRE

Control Action		Processes (Bubbles)				Output Control	
		PLC SRE on Pallet & Pick from Buffer	PLC SRE on Pallet & Pick from Line	PLC SRE in Buffer & Pick from Line	Maintain Buffer Count		
Pick_SRE _GO	PLC_SRE_ GO	1	2	3	4	Pick_SRE _DONE	PLC_SRE _DONE
"False"	"False"	0	0	0	1	"False"	"False"
"False"	"True"	1	0	0	1	"False"	"True"
"True"	"False"	0	0	1	1	"True"	"False"
"True"	"True"	0	1	0	1	"True"	"True"

TABLE 6.3. PAT for Transfer SRE

pointing and clicking the mouse on the C-Spec while operating in the Teamwork environment. The syntax of a PAT is as follows:

1. The PAT's top row (left to right)
   - Names of the control actions (input flows)
   - Numbers of processes (bubbles) on the accompanying DFD/CFD
   - Names of output controls (output flows)
2. Marked cells:
   - Input values, e.g., "False" or "True" in the Control Action columns.
   - Activation numbers, e.g., 0 or 1 in the Processes (Bubbles) columns.
   - Output values, e.g., "False" or "True" in the Output Control columns.

In Figure 7, Bubble 1 represents the action to be taken if the transfer process receives a "PLC_SRE_GO" signal; Bubble 2 represents the action to be taken when both "PLC_SRE_GO" and "Pick_SRE_GO" signals are received simultaneously; Bubble 3 defines the action taken when only a "Pick_SRE_GO" signal is received. This logic is specified in Table 3. An activation number of 1 means that the process is enabled, and a zero activation number means that it is disabled. The last three rows in Table 3 also demonstrate the concurrency of Bubble 4 with other Bubbles.

- Level 4:
  A detailed specification of the process "PLC SRE on Pallet & Pick from Line", is illustrated by its child process in Figure 8. The process consists of two elemental tasks; the first of which places the SRE subassembly on the DE pallet, and the second one picks a new SRE subassembly from the line. In

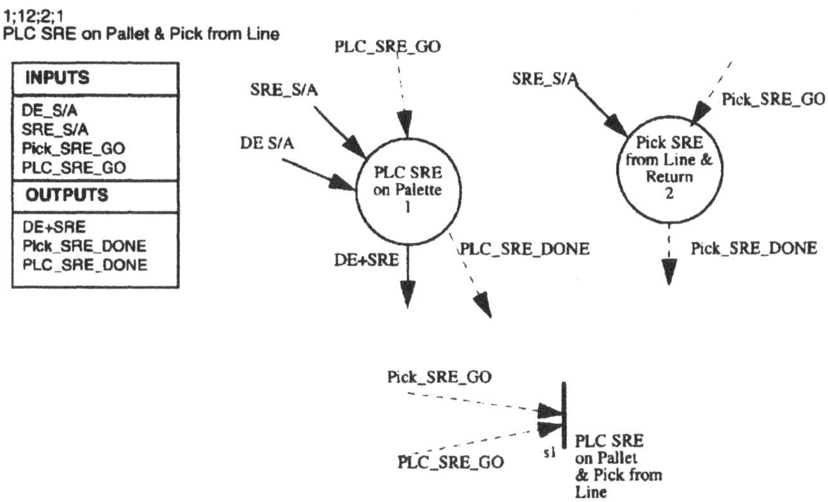

FIGURE 6.8. Specification of PLC SRE on Palette & Pick from Line (Level 4)

Control Action		Processes (Bubbles)		Output Control	
		PLC SRE on Pallet	Pick SRE from Line & Return		
Pick_SRE_GO	PLC_SRE_GO	1	2	Pick_SRE _DONE	PLC_SRE _DONE
"True"	"True"	1	2	"True"	"True"

TABLE 6.4. PAT for PLC SRE on Pallet & Pick from Line

addition, Table 4 describes the control logic that determines the activation and sequencing of these two operations, along with the value of the output control signals that are generated for each combination of input signals.

Hierarchical decomposition of processes through DFD/CFD and control specification expressed through PATs ensure that all possible situations are explicitly defined and able to be examined. The P-Specs for the processes "PLC SRE on Pallet" is shown in Figure 9. By specifying the process and control logic through these representations (DFD/CFD, PAT, P-Spec), varying levels of details can be created, changed, and maintained.

Modular Petri Net Representation

The functional partitioning of processes and tasks enables the modular examination of a specific process only (e.g., "Transfer SRE"), without any regard to other

NAME:
1.12.2.1;1

TITLE:
PLC SRE on Pallet

INPUT/OUTPUT:
DE_S/A: data_in
SRE_S/A: data_in
DE+SRE: data_out
PLC_SRE_GO: control_in
PLC_SRE_DONE: control_out

BODY:
```
/* This procedure places the SRE (in the robot's grasp) on the
 * DE pallet after it receives a PLC_SRE_GO signal
*/
IF PLC_SRE_GO = "True"
   Place SRE on pallet
   PLC_SRE_DONE = "True"
END IF
```

FIGURE 6.9. P-Spec for "PLC SRE on Pallet"

assembly processes or the control operations. All changes that are made remain in the domain of the given process. This modularity provides for flexibility in its initial specification or subsequent modification, since the information is easily retrieved, traced, modified, and separated.

The DFD/CFD specification of the transfer operation does not define the *timing* of inputs and outputs, nor does it check the reachable state space, given an initial state. This information can be graphically specified and verified with Petri Nets. The Petri Net specification, therefore, complements the DFD/CFD specification to form a comprehensive description of the system logic.

In order to remain consistent with the Teamwork specification and provide a concurrent description of functional and real-time logic at all levels of detail, the processes are partitioned identically in the P-NUT specification. Each of the possible actions taken by the transfer robot is represented as a sub-net. These sub-net are then combined through enabling transitions to form the complete Petri Net specification of the transfer operation, similar to the combination of processes in the Teamwork specification. The enabling transitions represent the same control logic as controlled in the PAT of Table 3, but in a visual form that is more representative of actual process control. The sub-nets that are used to represent the operations of "Transfer SRE" are:

1. Sub-net #1 - PLC SRE on Pallet & Pick from Buffer
2. Sub-net #2 - PLC SRE on Pallet & Pick from Line

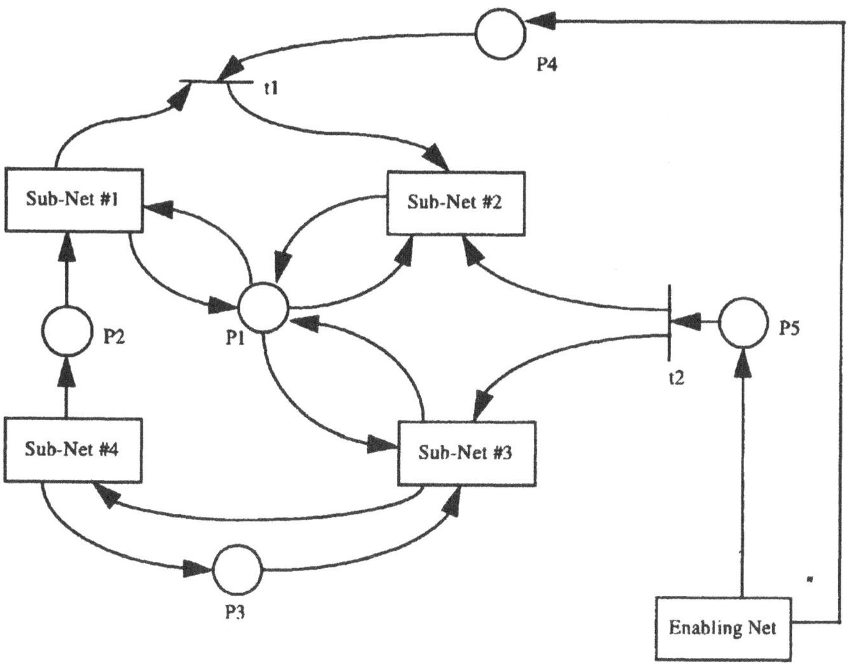

FIGURE 6.10. Modular Representation of Petri Net for Transfer SRE

3. Sub-net #3 - PLC SRE in Buffer & Pick from Line
4. Sub-net #4 - Maintain Buffer Count

The modular representation of the Petri Net specification provides a definition of the real-time logic for the DFD/CFD of Figure 7. The arrival of DE pallets and SRE pallets are the key events that enable the entire transfer process. These two events are combined into a single sub-net called an *enabling net*. The enabling net contains the initial markings (of places) that represent the initial state of the system. The connection of all four sub-nets and enabling net is shown in Figure 10.

Internal Specification of Sub-Nets

The internal specification of each sub-net describes the actions and conditions necessary to complete each process. An internal specification provides a description of the elemental tasks much like the child diagram for each of the bubbles in the Teamwork specification of Figure 7.

The graphical representation of the Petri Net specification for the transfer process shows the interconnections between sub-nets and the enabling net (Figure 11). The partitioning of processes and the detailed specification of each sub-net is traceable to the modular representation of Figure 10.

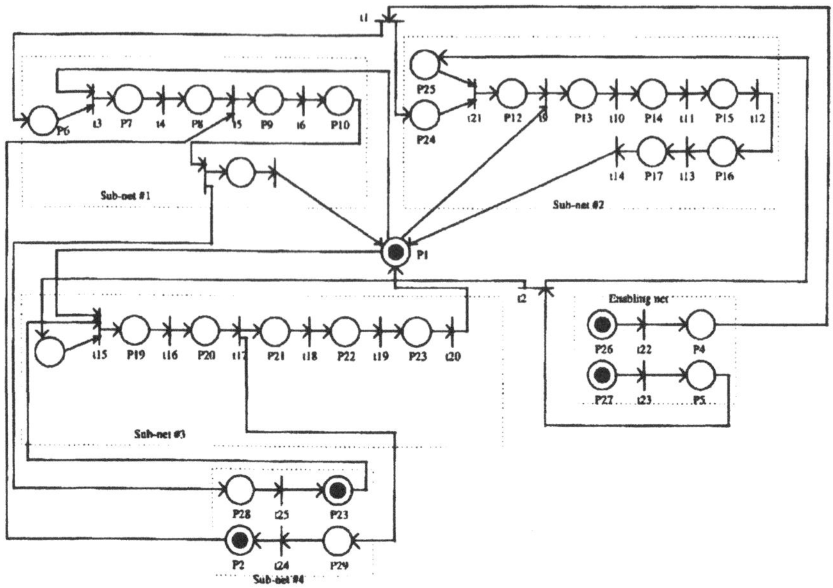

FIGURE 6.11. Graphical Petri Net Representation of Transfer SRE

In Figure 11, each sub-net specification is enclosed within a dashed line. Solid lines crossing these boundaries represent net inputs and outputs of the nets. The marking of places, P1, P2, P3, P26, and P27, represent one of several possible initial states for the system. Thus the Petri Net representation can specify and illustrate the process logic and the states of the system on the same diagram.

The strength of the Petri Net specification lies in its ability to aid in the visual identification of conflicting states and actions. Examination of Figure 11 reveals that transition t3, t9, and t15 are in conflict. They all require place P1 to be marked as an input. All these transitions will not be able to fire simultaneously, since the firing of any one of these transitions will cause the token place P1 to be moved, thereby disabling the remaining two transitions. Transition t3 and t9 describe the same actions taken by the transfer robot (Transfer robot begins placing SRE S/A on pallet), but are performed during separate modes of operations. Therefore, even though these two transitions are identical in terms of actions, they must be in conflict to ensure that only one mode of operation is enabled or active at any time. In addition, the sequential occurrence of events and transition of states within each sub-net is also clearly illustrated in the Petri Net specification. For instance, in Sub-net #1 of Figure 11, the precedence of P7 over P8 is visually defined, together with t4 necessary for the transition between these states.

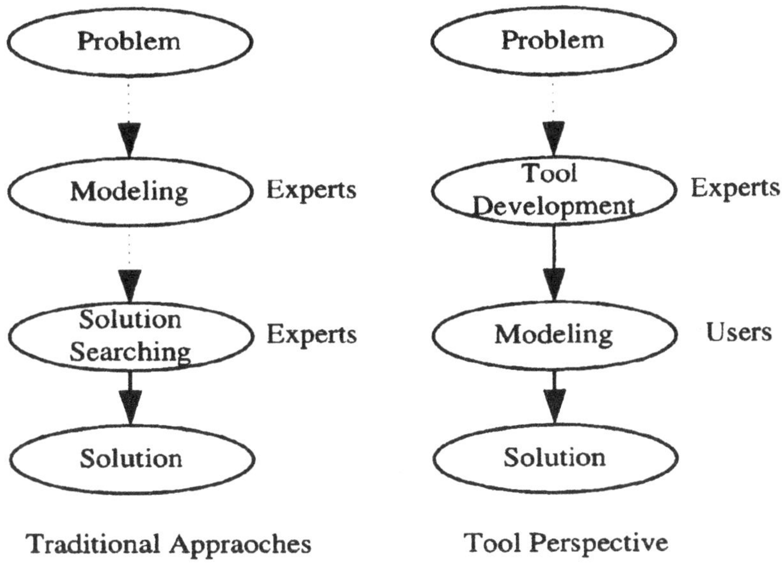

FIGURE 6.12. A Comparison of Traditional Approaches and Tool Perspective

6.4.4 Flexibility of Specification

The conflict problem described in the previous section can be resolved easily since the CFS model is developed on Teamwork and P-NUT. Both tools are computerized and can be modified like the editing function provided by general computer software. Nevertheless, the modification in both tools update not only the problem node or arc, but also the nodes and arcs which are related with the problem node or arc. Thus, the flexibility of specification is provided by the flexible changing capability.

6.5 Discussion and Conclusion

The impacts of computer and communication technologies have changed the focus of models of material handling and robotics (MHR) from "what are the recommended solutions?" to "how can the users resolve and implement the recommended solution?" For "what is the recommended solution", there must be experts who can formulate the problems and then search for the solutions. However, for "how can the users resolve and implement the recommended solution", the designers and users can develop models and resolve the problems by themselves based on the given *tools*. This tool perspective is one of the initial ideas of developing FDL and CFS for MHR models. Figure 12 shows the comparison of the two approaches. The dot arrow lines represent the difficulty to go through from one state to the other. For traditional approaches, the difficulty occurs from problem to modeling and from modeling to solution searching. For the tool perspective, the difficulty occurs only in developing the tools.

In this chapter, this concept is realized in two models: *facility description language* and *concurrent flexible specifications*. The two models address the concerns of traditional approaches. FDL functions through syntax in a 3-D emulation environment. Hence, users can check and modify the results of their models directly. In the models of concurrent flexible specifications (CFS), the users can develop the physical layout, functional model, information model, and control signals in an integrated modeling environment. In addition, the simplicity of validation and modification of users' design is another contribution of CFS in the modeling of material handling and robotics.

It is anticipated that integrated modeling approaches with a tool perspective will continue to evolve. Interesting challenges include the development of learning mechanisms and computer-supported collaborative modeling tools.

Acknowledgment

Research reported in this chapter has been developed in the PRISM Program at Purdue University with particular support from the NSF Grants DDM92-14143 and DDM93-09579, "Models for Design of Engineering Task Integration."

6.6 REFERENCES

[1] Alabiso, B., "Transformation of Data Flow Diagram Analysis Models to Object Oriented Design," *OOPSLA '88 Proceedings*, pp. 335-353, 1988.

[2] Askin, R. G. and Standridge, C. R., *Modeling and Analysis of Manufacturing Systems*, John Wiley & Sons, 1993.

[3] Cayenne Software Inc. *TeamWork for Structured Methods*, http://www.cayennesoft.com/teamwork/, 1997.

[4] Chadha, B., Fulton, R. E., Calhoun, J. C., "Design and Implementation of a CIM Information System," *Engineering with Computers*, Vol. 10, No. 1, pp. 1-10, 1994.

[5] Csurgai, G., Kovacs, V., and Laufer, J., "A Generalized Model for Control and Supervision of Unmanned Manufacturing Cells," *Software for Discrete Manufacturing*, IFIP, pp. 103-112, 1986.

[6] Furtado, G. P. and Nof, S. Y., "Concurrent Flexible Specifications of Functional and Real-Time Logic in Manufacturing," Research Memorandum No. 95-1, School of Industrial Engineering, Purdue University, 1995.

[7] Furtado, G. P., Huang, C. Y., and Nof, S. Y., "Concurrent Flexible Specifications of Functional and Real-Time Logic in Manufacturing," *IEEE Transactions on Engineering Management*, Forthcoming, 1998.

[8] Hitomi, K. and Yoshimura, M., "Operations Scheduling for Work Transportation by Industrial Robots in Automated Manufacturing Systems," *Robotics and Material Flow, S. Y. Nof (ed.)*, Elsevier, pp. 131-139, 1986.

[9] Nadoli, G. and Rangaswami, M., "An Integrated Modeling Methodology for Material Handling Systems Design," *Proceedings of the 1993 Winter Simulation Conference*, pp. 785-789, 1993.

[10] Nof, S. Y., "On the Structure and Logic of Typical Material Flow Systems," *Int. J. of Production Research*, Vol. 20, No. 5, pp. 575-590, Sep.-Oct., 1982.

[11] Nof, S. Y. and Lechtman, H., "Robot Time and Motion," *Industrial Engineering*, pp. 38-48, April 1982.

[12] Nof, S. Y. and Drezner, Z., "Part Flow in the Robotic Assembly Plan Problem," *Robotics and Material Flow*, S. Y. Nof (ed.), Elsevier, pp. 197-205, 1986.

[13] Prakash, A. and Chen, Mingyuan, "A Simulation Study of Flexible Manufacturing Systems," *Computers in Industrial Engineering*, Vol. 28, No. 1, pp. 191-199, 1995.

[14] Razouk, R. R., "A Guided Tour of P-NUT (Release 2.2)", ICS-TR-86-25, Department of Information and Computer Science at the University of California, Irvine, January 1987.

[15] Sinriech, D., "Network Design Models for Discrete Material Flow Systems: A Literature Review," *Int. J. Advanced Manufacturing Technology*, Vol. 10, pp. 277-291, 1995.

[16] Sly, D. P., "Plant Design for Efficiency Using AutoCAD and FactoryFLOW," *Proceedings of the 1995 Winter Simulation Conference*, pp. 437-444, 1995.

[17] Suri, R., "Quantitative Techniques for Robotic Systems Analysis," *Handbook of Industrial Robotics, S. Y. Nof (ed.)*, John Wiley & Sons, N. Y., pp. 605-638, 1985.

[18] *Tecnomatix Technologies, ROBCAD TDL Reference Manual*, version 2.2, December 1989.

[19] Witzerman, J. P., and Nof, S. Y., "Facility Description Language (version 1.1)–User Manual," Research Memorandum No. 95-5, School of Industrial Engineering, Purdue University, May 1995a.

[20] Witzerman, J. P., and Nof, S. Y., "Facility Description Language," *Proceedings of IERC4*, Nashville, TN, pp. 449-455, 1995b.

[21] Witzerman, J. P., and Nof, S. Y., "Integration of Simulation and Emulation with Graphical Design for the Development of Cell Control Programs," *Int. J. of Production Research*, Vol. 33, No. 11, pp. 3193-3206, 1996.

7

A Simultaneous Approach for IMS Design: a Possibility Based Approach

G. Perrone[1]
S. Noto La Diega[2]

ABSTRACT IMS investments are characterised by high fixed costs and long life cycles. On the other hand, their redditivity and risk coverage depend on their manufacturing efficiency that is mainly defined during the design phase by fixing the system configuration. Due to the flexibility required to IMS, system configuration depends not only from technological information, such as product routing table and service times, but also from marketing data such as the typology of products to be manufacture and their production volumes. Moreover, the evaluation of the redditivity and the risk of the investment depends on market information such as product prices as well. Such interdependencies make the investment decision environment very complex. Furthermore, the requirements and the data characterising the decision environment are affected by imprecision due to the strategic nature of the decision. Therefore, IMS investment decisions have to be made in a very complex and vague decision environment. Even if, several approaches have been proposed in literature to deal with IMS design problem, very few consider the specific characteristics of the decision environment. In this paper we propose a methodology and a tool that fit very well with the complexity and the vagueness of such design problem. The methodology consists into a simultaneous description of all the requirements, while the tool is the fuzzy possibility theory.

7.1 Introduction

Flexibility is acquiring more and more relevance in manufacturing strategy because it allows:

- to cope the uncertainty that results from changes or fluctuations in level of demand, production price, production mix and action of competitors [13];
- to cope the disturbances in internal environment as equipment breakdowns, variable task times, queuing delays, rejects and reworks [3];
- to make the manufacturing process both cost efficient and effective in producing customised products without sacrificing other objectives, through the

[1]DIFA - University of Basilicata, Potenza - Italy
[2]DTPM - University of Palermo - Italy

emphasis on the scope [15] rather than on scale economies;

- to undertake both offensive and defensive strategies [41].

It is not surprising that, massive investments are being made in new advanced manufacturing technologies incorporating high flexibility such as Integrated Manufacturing Systems (IMSs). Such manufacturing systems are characterised by high level of investment and a consequently by high level of risk [26].

This paper deals with the IMS design problem. There exists an extensive body of literature on the multitude of developed techniques for the IMS design. A brief summary of these methods is: queuing network theory [4, 6, 37, 38], [40], simulation models [5], multiple criteria analysis models [39], mathematical programming models [2, 28].

This brief review shows that there exist proven methods for demonstrating the technical feasibility of the designing IMS, but all of them consider IMS design decisions imposed by market strategy and financial constraints. Indeed, the general approach to IMS design can be summarised as it follows: "*first the company fixes the market strategy including products definition and competitive characteristics (volumes and prices), then the IMS will be designed in order to accomplish the previous decisions and to meet the financial requirements*".

This approach is in contrast with the importance the manufacturing has acquired during the last decade. Nowadays it is fully acknowledged, that manufacturing plays a strategic role in the corporate strategy. Many researchers have highlighted that manufacturing has to go beyond simple alignment with the market policy. In fact, it has to make a positive contribution to the firm's competitive position [20, 30] making the company able to react to the market, to align with the market and even to help in creating the market. This can be done focusing on flexibility [36, 42]. But, when flexible manufacturing systems are implemented, their performances are not only depending on single products, but on the product mix to be manufactured. In this sense, market characteristics have strong relevance on manufacturing performances. In fact, product typology and volumes may influence the degree of efficiency of the manufacturing systems impacting the complexity of the material flow within the system. Vice versa, since production efficiency affects the strength of the economies obtainable, manufacturing performances can improve market competitiveness. Since modern flexible manufacturing systems are characterised by high fixed cost, the higher is the value of the scope and scale economies obtainable, the lower is the product unit cost. Having lower production cost allows to gain high margins and therefore to improve the competition strength of the firm.

For the above reasons the classical approach to IMS design can be considered incorrect. This is not due to the design method used, but to the hierarchical approach used to address the problem that postpones manufacturing decisions to marketing and financial ones. In other papers, we have stressed how marketing and financial consideration can be used to correctly design the amount of flexibility to put into a IMS [23, 32]. Here an approach to design effective and efficient IMS that simultaneously consider market, finance and manufacturing requirements is proposed.

However, since market decisions are essentially strategic, an approach for simultaneously consider market, financial and manufacturing requirements in IMS design has to involve strategic considerations so that we can refer to it such as "*Strategic IMS design*" approach.

In this paper we are going to formalise from the conceptual and operative point of view the "*Strategic IMS design*". In section 7.2 the decision making reference environment is analysed. Being strategic decision essentially human processes, they are affected by heavy vagueness, therefore, in section 7.3, we advocate the possibility theory as tool to make decisions in the environment pointed out in section 7.2. In section 7.4 an attempt framework for strategic IMS design is proposed: this framework formalise in mathematical way the decision environment discussed in section 7.2 using the tool advocated in section 7.3. In section 7.5 a brief example is presented in order to highlight the proposed methodology. Finally, in section 7.6 conclusions of this research are discussed.

7.2 The decisional environment for Strategic IMS Design

The initial, and perhaps the most difficult, decisions to be made when a strategic IMS design is faced, concern the technological and the production capacities. The *technological capacity* deals with the number of different technological operations the manufacturing system is able to perform. It is straightforward that, the higher is the number of the different technological operations the manufacturing system is able to perform, the wider is the number of different products simultaneously workable by the manufacturing systems. The *production capacity* deals with the number of resources for each technological resource. This number depends on production volumes of each technological operation and, considering the whole manufacturing system, on the aggregate volume.

Another very important design variable to be settled during this phase is the *configuration of the manufacturing systems*, that is to say how to organise the production flow within the manufacturing system. The manufacturing system configuration depends essentially on the machine flexibility. The extreme hypotheses are a pure dedicated resources configuration and a pure general purpose one. When a *pure dedicated resources configuration* is selected, each workstation has the lowest flexibility being able to perform just a single technological operation, or a group of similar technological operations. In this case the production flow, within the manufacturing system, is comparable to a pure job-shop, each product has a fixed routing and, as result, the efficiency of the manufacturing system is low.

On the other hand when a *pure general-purpose configuration* is chosen each workstation has the highest machine flexibility being able to perform a wide range of technological operations. As a result of such machine flexibility, the production flow is comparable to a system of parallel identical machines, each product has several routing alternatives and, as result, the expected efficiency of the manufacturing system is higher. Of course the former configuration is less expensive then the latter.

The definition of the above variables depends on the definition of the firm market strategy. Particularly, it depends on what the firms intends to manufacture, how it would like to operate and what kind of requirements and constraints the company has to face. This defines the decision environment in which the manufacturing decisions are to be made. That is to say that, the design of a flexible manufacturing system

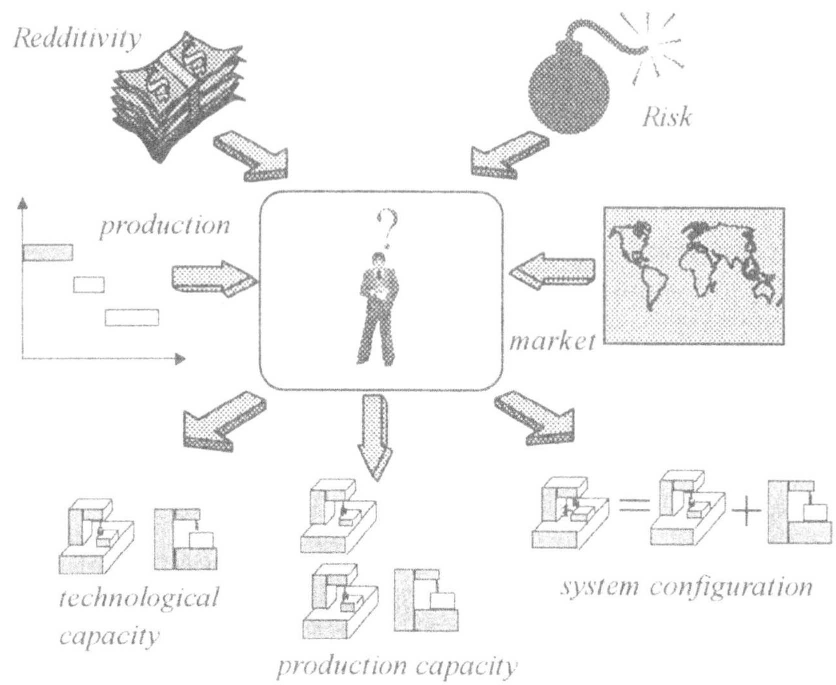

FIGURE 7.1. Decision environment for strategic IMS design

depends on the environment in which it has to work.

A decision tool for a flexible manufacturing systems design must be able to consider all the requirements and constraints such an environment introduces on the decision process. In our opinion the environment characterising such a decision is depicted in Figure 7.1. In it, is possible to distinguish four main issues: market, production, redditivity and risk. Each one of these topics introduces different requirements and constraints on the design decision. Let us examine each topic one at the time.

Market
Market requirements and constraints concern what kind of competition the company intends to play by using the manufacturing system under design. Since the market defines the set of products that the firms should be able to manufacture, the company *marketing strategy* has to be examined firstly. The company could pursue a niche strategy, deciding to produce some of the products belonging to that market, or to produce all the products in the market pursuing a differentiation strategy, or, finally, to produce just few products pursuing a cost strategy [34]. In any case, such a decision, i.e. what kind of products to manufacture, influences the technological capacity of the manufacturing system.

Afterwards, for each product, the *product marketing strategy* has to be analysed. This strategy concerns the competitive position the firm wants to obtain within the market. Production volumes and prices settle the competitive position for each product, but their definition is constrained by the actual market competition and economic situation. Therefore, both macro and micro economic considerations have to be made in order to define competitive position of each product, but in any case, such choice influences both technological and production capacity definition.

Finally, the *mix strategy* has to be considered. This issue concerns the relation among products due to aggregate production strategy or physical links among the products. Such relation imposes some constraints on the production ratios influencing the production capacity definition.

Production
Production has often a conflict with the marketing regarding the selection of those products to be manufactured. In fact, production looks for manufacturing systems efficiency while marketing aims to gain good market positions, therefore, they often originate a dichotomy that is the origin of the so called make or buy decisions. Such a dichotomy influences the manufacturing system design, since production requirements and constraints concern directly the configuration and indirectly the technological and production capacity.

Production efficiency aims for the simplification of the production flow and therefore for higher manufacturing resources utilisation. This pushes towards general-purpose configuration. The possibility to obtain good general-purpose configuration depends on the technical feasibility to group technological operations together and, therefore on the typology of the technological operations required by the products to be manufactured. For the above reasons, production efficiency indirectly asks for products whose technological operations are easily grouped in general purpose equipment. By influencing product selection, the production affects the technological capacity.

Technical feasibility is a necessary issue to make a choice between dedicated or general- purpose configuration, but it is not a sufficient one. Also economic considerations are to be taken into account. Flexible resources, indeed, are most expensive than dedicated ones, so in order to be economically convenient they have to be better utilised. Workstation utilisation basically depends on the resources aggregate workload, therefore, production efficiency indirectly influences also production volumes.

Finally, production has to consider the capacity constraint, i.e. the resource workload must be lower then production capacity available. This constraint is influenced by the configuration, by the mix of the products and by their production volumes.

Redditivity
Redditivity involves the financial requirements of the company. The financial department puts some requirements and constraint concerning both the available budget for the project, but, overall, the expected economic return of the manufacturing investment in IMS. Several economic measures can be utilised to express this last requirement such as the Net Present Value, the Internal Return Revenue, the Return on Investment index and so forth. Since both investment amount and redditivity are influenced by the products to be manufactured and by their production volumes and prices, they influence the technological and production capacity. Moreover, since manufacturing system efficiency can improve the system performances, the redditivity requirements affect the system configuration as well.

Risk
Risk deals with the entrepreneur risk propensity, that is the capacity of each decision-

	Requirements	*Technological Capacity*	*Production Capacity*	*System Configuration*
Market	*Market Strategy*	√		
	Product Strategy	√	√	
	Mix Strategy		√	
Production	*System Efficiency*	√	√	√
	Capacity Constraint	√	√	√
Redditivity	*Economic Return*	√	√	√
Risk	*Investment Risk*	√	√	√

TABLE 7.1. Decision environment requirements and constraints

maker to accept a given amount of risk in each investment. For any kind of instrumental investment, risk can be measured by methods such as the Pay Back Period. A risk requirement can be formalised as the amount of risk the decision-maker can accept in order to undertake the investment project. Any measure of the risk involves the decision variables of the problem.

As the reader can notice, a *Strategic IMS Design* problem can be related as a decision problem in which each company department expresses its requirements. Such requirements involve the same design parameters. Often those requirements are conflicting each other, always they interact each other. Table 7.1 summarises the discussed interaction among the requirements. As result, a *Strategic IMS Design* problem is a complex multiple objective decision problem.

7.3 The Strategic IMS Design Decision-Making Tool: the possibilistic programming theory

A decision making tool for Strategic IMS Design must be able to formalise all the requirements and constraints the described environment places on such a decision. In order to understand what kind of formalisation can be utilised for our purposes, it is necessary to analyse how each requirement and constraint is expressed in the actual decision environment. It is straightforward that the decision environment discussed in the previous paragraph is a strategic one. Strategic decision environment are characterised by:

- *Vagueness.* Vagueness is a kind of uncertainty due to the imprecise definition of the decision environment. In our framework every requirement, constraint and goal has a strategic definition. A strategic assessment is the result of a human decision process and for this reason it cannot be expressed by a mathematical formalisation, but through a linguistic one. Linguistic categories are, indeed, instruments of human reasoning and each result of this process is a linguistic expression. Linguistic assessments are the main font of vagueness in decision-making problems, therefore, our formalisation tool must be able to deal with linguistic categories and to give a mathematical representation for them.

- *Approximation.* This problem is typical in strategic decisions and it deals with the attempt to foresee possible scenarios. This is a very complex task, above all, when the scenario is new or far from the actual situation. Several methods have been proposed to approach this problem. Surely, methods based on expert evaluations are widely used in industry. The results of such methods are often approximate estimations of the data characterising the decision environment; an approximate estimation is a kind of uncertainty related again with the low availability of information. Therefore, the decision making tool has be able to face uncertainty deriving from approximate estimation of data.
- *Complexity.* The framework gives an overview of such a complexity which mainly is due to the multiple objective aspect of the problem; each issue, indeed, puts different requirements and constraints on the decision and often those assessments are conflicting each other. The decision model has to be a multiple objective one and it must be able to compromise different and conflicting objectives.

Vagueness, data imprecision and conflicting requirements are the main characteristics of the framework where the IMS design decision is to be made as they are the main characteristic of a strategic decision environment. One the other hand, designing a manufacturing system is an engineering decision, i.e. a decision where numbers have to replace words. Therefore, we need to turn strategic assessments, which are fuzzy in nature, in engineering decisions which are essential crisp.

Several attempts have been made to address the imprecision problem in engineering decision problems in literature. Davis et al. [7], Kim et al. [18], Navinchandra et al. [27], Finch et al. [12], have attempted to use interval mathematics to address the imprecision problem in product development. These approaches have had limited success. Other approaches used to address imprecision in designing problem include utility theory, implicit representation using optimisation methods, matrix methods and probability methods. Reusch [35] examined the general problem of imprecision and inconsistency in design and also concluded that the problems are well suited to be solved using fuzzy mathematics. Antonsson [1] made an extensive review and comparison of these methods applied to modelling imprecision in design and concluded that only fuzzy set theory is satisfactory. Dubois, [10], and Trappey [44], pioneered the application of fuzzy set theory to express requirements in engineering problems. Wood et al. [45], and Otto et al. [29], applied fuzzy set theory to preliminary engineering design and to product design using a variation of Dong and Wong's [8] discretized solution algorithm in a method related to Taguchi's method. However, their approach has computational complexity of the order $O(M \cdot 2^{N-1} \cdot k)$, where N is the number of imprecise parameters, M is the number of α - levels into which the membership function is divided, and k is the number of multiplication and divisions in the function being evaluated. Examining this work we see that only small, isolated pieces of design problems have been investigated and that fuzzy mathematics has not been applied to all design phases. Young et al. [46], and Giachetti et al. [14], have shown the suitability of Fuzzy Constrained Networks to model design problem affected by linguistic and imprecision vagueness. Their approach consists in turning a set of imprecise requirements into a network of numerical constraints by using fuzzy mathematics. Afterwards the network is solved through a propagation mechanism of the truth based on fuzzy logic. They have shown how this method allow to model and solve engineering design problem, but they have also stressed how the truth propagation mechanism does not assure either to find optimal or compromise solutions. Finally, Perrone et al. [33], have shown how possibility theory (see [17]) can be successfully used to turn linguistic requirements into mathematical models to be optimised by using evolutionary techniques.

Using possibilistic formalisation, indeed, is possible to give a mathematical representation of the discussed vagueness and imprecision and to turn the decision problem into a possibilistic multiple objective programming. In fact, vague requirements and/or constraints can be easily turned into fuzzy sets expressing the degree of satisfaction of the requirement when the decision variable varies in its domain. Likely, imprecise data can be formalised through fuzzy numbers expressing the approximate knowledge of the data. In such an environment also decision variables are consequently expressed by using fuzzy set. In particular, four relative modalities have been defined in Possibility Theory to address a requirement satisfaction given a value of the decision variable, but possibility and necessity are the most used. The *possibility modality* measures the degree of truth of an uncertain data (ξ) to meet a vague requirement, constraint and/or goal (μ). This is performed through the following superior/minimum measure corresponding to a max/min decision-making approach:

$$\pi(\mu/\xi) = \sup_x(\min(\mu(x), \xi(x)) \tag{7.1}$$

Figure 2a gives the possibilistic formalisation of the vague requirement "*the uncertain variable ξ should vary within the range about [a,b]* ". ξ is the fuzzy number representing the uncertainty on the decision variable ξ, and μ is the fuzzy set expressing the degree of decision maker satisfaction when ξ varies within $[a, b]$; the possibility measures, $\pi(\mu/\xi)$, of Figure 7.2a can be interpreted as the degree of satisfaction of the above vague requirement. Similarly, Figure 7.2b gives the possibilistic formalisation of the constraint "*the uncertain variable ξ should be greater then approximately z*". Also in this case, the possibility measures, $\pi(\mu/\xi)$, of Figure 7.2b can be interpreted as the degree of satisfaction of the above vague constraint.

Possibility theory can be used to formalise vague objective as well. Let us suppose that a function $f(x)$ needs to be maximised. Decision-maker has often an idea of what can be fully satisfying and what is completely unsatisfactory. This means that, it is always possible to set a range $[a, b]$ such that, in $]-\infty, a]$ the decision-maker is completely unsatisfied, in the range $[b, +\infty[$ she/he is completely satisfied while in the range $]a, b[$ her/his satisfaction is increasing. Such an objective can be formalised by using the same representation of Figure 7.2b.

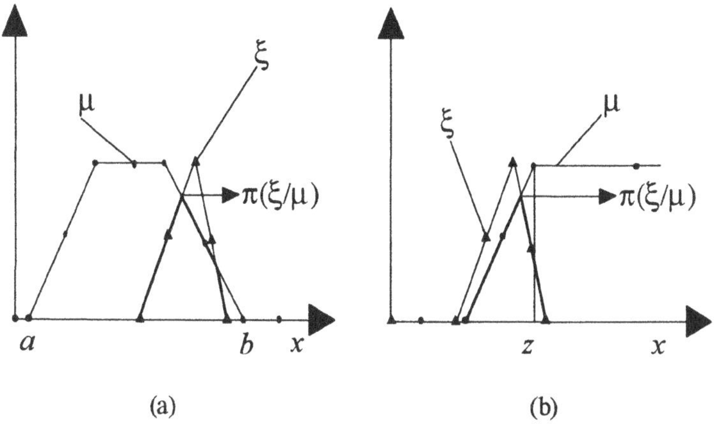

FIGURE 7.2. Possibility measure

A very important issue using possibility theory is that also implication requirements, i.e. IF - THEN relations, which are very widely used in human language, can be formalised into a mathematical way using fuzzy inferences techniques [11]. This can be done by modelling linguistic requirements by means of linguistic variables and applying relative modalities and fuzzy implication operators to address the degree of truth of the linguistic requirements.

Moreover, using possibility theory is also possible to stress critical requirements, i.e. those requirements the decision-maker considers of very high importance. This can be accomplished by using the necessity modality instead of the possibility one. Possibility gives a measure of the degree of truth of a fuzzy requirement μ given a fuzzy event ξ, but it does not say anything about what does happen to the fuzzy event under the opposite requirement, $(\neg\mu)$. Necessity modality, instead, gives a measure of the degree of impossibility of the requirement $(\neg\mu)$ when the event happens, i.e.:

$$\eta(\mu/\xi) = [1 - \pi(\neg\mu(x)/\xi(x))] = \inf_x(\max(\mu(x), \neg\xi(x)) \tag{7.2}$$

Therefore, it is a more pessimistic or conservative evaluation and it can be applied to appoint such requirements that more then other need to be satisfied. Figure 7.3a and 7.3b show the necessity modality when it is applied at the same situation of figure 7.2a and 7.2b.

Let S be the set of all the requirements, constraints and goals the decision environment of Figure 1 puts on the IMS design decision and $\nu_{x\in X}$ the set of decision variables of the problem. Let also $\delta_S(\nu_1, \ldots\nu_X)$ be the degree of satisfaction of each s in the set S, being obtained using either possibility or necessity modalities or a fuzzy inference.

The decision-maker aims at maximising the degree of true of each s, so that the decision can be formalised as it follows:

$$\max\left(\bigcap_s \delta_s\right) \tag{7.3}$$

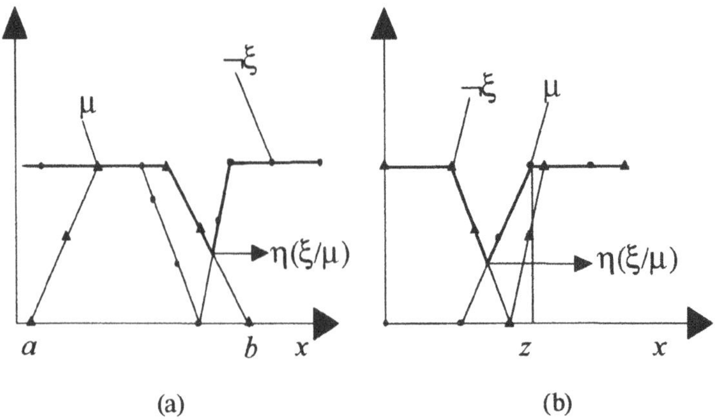

FIGURE 7.3. Necessity measure

By using the minimum operator to perform the above intersection, the decision problem (7.3) becomes:

$$\max\left(\bigcap_{s}\delta_s\right) = \max\left[(\min_{S}(\delta_s)\right] \tag{7.4}$$

that can be turned into the following programming model:

$$\max(\lambda_{\min}) \tag{7.5}$$

$$s.t.$$

$$\lambda_{\min} \leq \delta_s, \forall s$$

$$\textit{domain constraints of each } \nu_x, \textit{ with } \mathrm{x} \in X$$

The formulation of (7.5) is suitable to figure out compromise solutions, but not non-dominated ones. In order to obtain non-dominated compromise solutions [22] the following programming model needs to be considered:

$$\max(\lambda_{\min} + \lambda_{avg}) \tag{7.6}$$

$$s.t.$$

$$\lambda_{\min} \leq \delta_s, \quad \forall s$$

$$\lambda_{avg} = \frac{\sum_s \delta_s}{S}$$

$$\text{domain constraints of each } \nu_x,\ x \in X$$

The solution of model (7.6) is quite complex. Very often, indeed, the relations linking the δ_s with the decision variables are not linear. Moreover, the decision variables themselves can be real, integer or dummy. For these reasons, no classical linear programming solver can be used to solve problems such as (7.6). The optimisation of the model can be performed by using non-linear optimisation techniques like random approaches, or genetic algorithms. These techniques are not able to lead to optimal solution. It is to be stressed, however, that in such strategic decision problems the decision-maker looks for satisfying solutions of the problem and not for the best ones. The proposed model itself has been derived following this philosophy so that the research of the best solution will misrepresent the spirit of the method.

7.4 The possibilistic framework for strategic FMS design

In section 7.2 the decision environment for the IMS design decision has been discussed, while in the previous paragraph the possibilistic approach has been advocated for giving a mathematical formalisation of such decisions. In the present section the decision environment and the mathematical tool have been linked together obtaining an attempt framework that could be useful for making IMS strategic design decisions.

The proposed framework concerns a company entering in a new market by planning the accomplishment of a proper manufacturing system. Therefore, the proposed framework can be referred as the "*New entering strategic IMS design framework*". The proposed framework aims at being as much general as possible, but nevertheless, some particular hypotheses have been considered necessary in order to improve reader comprehension. However, such hypotheses do not reduce the generality of the proposed framework that can be considered as a reference one in case of a new entering case.

7.4.1 Market

Market influences products choice and therefore the typology of technological operations needed to manufacture those products, briefly market decisions directly impact technology capacity.

Marketing strategy

The marketing strategy can be formalised saying that the company aims to include not less then I^* products in its portfolio. Such a representation is quite general, since it can lead to a niche or a cost strategy when I^* is low or just one, to a differentiation strategy when I^* is high. Such concept can be formalised with the fuzzy set in Figure 7.4a.

Product strategy

The product strategy concerns some requirements the marketing has estimated for production volumes and prices. All those requirements are expressed through vague expressions such as: "*the production volume of the product i has to be about V_b units*", or "*the price of the product i has to be about p Currency Units (CU)*". Such vague

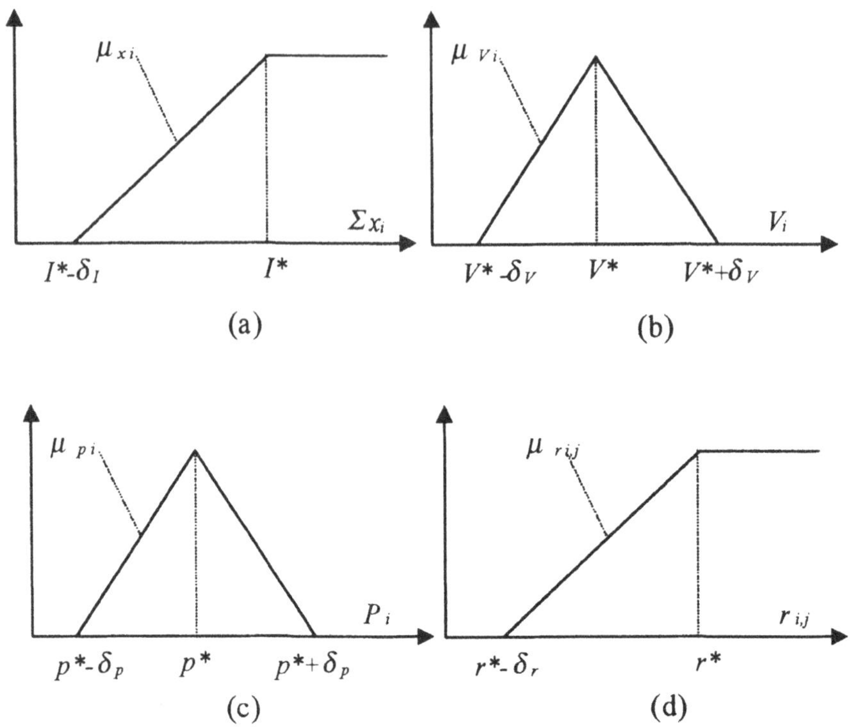

FIGURE 7.4. (a) Marketing strategy requirement. (b) Product strategy requirement: volumes. (c) Product strategy requirement: prices. (d) Mix strategy requirement.

requirements and constraints can be formalised using the fuzzy sets depicted in Figures 7.4b-c.

Mix strategy
Some vague requirements such as: "*the production ratio between the volumes of the products i and j, should be not less than r*" is used to highlight the relations linking couple of products. The fuzzy set of Figure 7.4d expresses such vague requirements.

Let us indicate with I the whole set of products within whom the company has to choose what to manufacture. From the market point of view, the problem decision variables are:

x_i, 1 if the product i is produced, 0 otherwise;
V_i, production volume of the product $i \in I$;
p_i, price of the product $i \in I$.

In our model, not dummy decision variables, such as prices and volumes, can be either crisp or fuzzy. Crisp variables are chosen when decision-makers aims at obtaining a precise evaluation of such variables satisfying the decision environment requirements. On the other hand, fuzzy variables are used when decision-makers prefer to have an approximate evaluation of such variables. This situation can be preferable when, in front of an imprecise knowledge of the decision environment, the decision-maker considers more useful to know a range where each decision variables can vary more or less satisfying the environment requirements. This approximation can be useful to locate the final crisp solution when more information about the decision environment are available.

In this last case decision variables can be represented through symmetrical triangular fuzzy numbers, where left and right spreads are equal to a fixed percentage, Δ, of the central value. In case of fuzzy decision variables the over sign "$\sim$" will be used to identify fuzzy numbers. For example, a fuzzy volume is denoted as $\tilde{V}_i$ and it stands for the triangular fuzzy number $[(1-\Delta)\cdot V_i, V_i, (1+\Delta)\cdot V_i]$.

By using possibility theory is now possible to formalise the satisfaction functions for each vague requirement of above. The mathematical formulation of the requirements of Figures 7.4a-d is given by the following expressions:

- *Degree of satisfaction of the marketing strategy requirement* (Figure 7.4a). In this case, being the decision variable a dummy one, possibility and necessity modalities give the same result, that is:

$$\pi(x_i)_{I^*} = \max\left(0; \min\left(\frac{(\sum_i x_i) - (I^* - \delta_I)}{\delta_I}; 1\right)\right) \tag{7.7}$$

  and, as the reader can notice, it depends on the decision variables x_i.

- *Degree of satisfaction of the volume requirements* (Figure 7.4b). Let us indicate with:

$$\xi_{\tilde{V}_i} = \max\left(0; \min\left(\frac{V_i\cdot(1+\Delta) - z}{\Delta\cdot V_i}; \frac{z - V_i\cdot(1-\Delta)}{\Delta\cdot V_i}\right)\right) \quad \forall i \tag{7.8}$$

  the membership function of the fuzzy set $\tilde{V}_i$ that is the actual volume of the product i; let also be

$$\neg\xi_{\tilde{V}_i} = \min\left(1; \max\left(\frac{V_i - z}{\Delta\cdot V_i}; \frac{z - V_i}{\Delta\cdot V_i}\right)\right) \quad \forall i \tag{7.9}$$

  the membership function of the opposite of $\tilde{V}_i$; finally, let be

$$\mu_{V_i} = \max\left(0; \min\left(\frac{z - (V_i^* - \delta_{V_i^*})}{\delta_{V_i^*}}; \frac{(V_i^* + \delta_{V_i^*}) - z}{\delta_{V_i^*}}\right)\right) \quad \forall i \tag{7.10}$$

  the membership function of the volume requirement of each product i. Therefore the satisfaction of the requirement (7.10) given the actual volume of (7.8) computed by (7.1) and (7.2) is respectively:

$$\pi_{V_i}(\tilde{V}_i) = \pi(\mu_{V_i}/\xi_{\tilde{V}_i}) \ \forall i \tag{7.11}$$

$$\eta_{V_i}(\tilde{V}_i) = \eta(\mu_{V_i}/\xi_{\tilde{V}_i}) \ \forall i \tag{7.11bis}$$

and, as the reader can notice, it depends on the decision variable $\tilde{V}_i$.

- *Degree of satisfaction of the price requirements* (Figure 7.4c). Let us indicate with:

$$\xi_{\tilde{p}_i} = \max\left(0; \min\left(\frac{p_i \cdot (1+\Delta) - z}{\Delta \cdot p_i}; \frac{z - p_i \cdot (1-\Delta)}{\Delta \cdot p_i}\right)\right) \ \forall i \tag{7.12}$$

the membership function of the fuzzy set $\tilde{p}_i$ that is the actual price of the product i; let also be

$$\neg\xi_{\tilde{p}_i} = \min\left(1; \max\left(\frac{p_i - z}{\Delta \cdot p_i}; \frac{z - p_i}{\Delta \cdot p_i}\right)\right) \ \forall i \tag{7.13}$$

the membership function of the opposite of $\tilde{p}_i$; finally, let be

$$\mu_{p_i} = \max\left(0; \min\left(\frac{z - (p_i^* - \delta_{V_i^*})}{\delta_{V_i^*}}; \frac{(p_i^* + \delta_{V_i^*}) - z}{\delta_{V_i^*}}\right)\right) \ \forall i \tag{7.14}$$

the membership function of the price requirement of each product i. Therefore the satisfaction of the requirement (7.14) given the actual price of (7.12) computed by (7.1) and (7.2) is respectively:

$$\pi_{p_i}(\tilde{p}_i) = \pi(\mu_{p_i}/\xi_{\tilde{p}_i}) \ \forall i \tag{7.15}$$

$$\eta_{p_i}(\tilde{p}_i) = \eta(\mu_{p_i}/\xi_{\tilde{p}_i}) \ \forall i \tag{7.15bis}$$

and, as the reader can notice, it depends on the decision variable $\tilde{p}_i$.

- *Degree of satisfaction of the volume ratio requirements* (Figure 4d). Let us indicate with:

$$\begin{aligned} \tilde{r}_{i,j} &= \left(r_{i,j} \cdot \frac{(1-\Delta)}{1+\Delta}, r_{i,j}, r_{i,j} \cdot \frac{(1+\Delta)}{1-\Delta}\right) \\ &= \left(r_{i,j} \cdot (1-\Delta_r^L), r_{i,j}, r_{i,j} \cdot (1+\Delta_r^R)\right) \ \forall i,j \end{aligned} \tag{7.16}$$

the fuzzy volume ratio obtained by dividing $\tilde{V}_i$ times $\tilde{V}_j$, and where:

$$\Delta_r^L = \frac{2 \cdot \Delta}{1+\Delta} \tag{7.17}$$

$$\Delta_r^R = \frac{2 \cdot \Delta}{1-\Delta} \tag{7.18}$$

Since fuzzy division is not conservative of the linear shape [9], $\tilde{r}_{i,j}$ has not a triangular shape. Nevertheless, according to [9], if the value of Δ is less than 10%, the error between the actual shape and the linear one is quite irrelevant, therefore, in this case, we can keep the triangular shape for $\tilde{r}_{i,j}$. Let also be:

$$\xi_{\tilde{r}_{i,j}} = \max\left(0; \min\left(\frac{r_{i,j} \cdot (1+\Delta_r^R) - z}{\Delta_r^R \cdot r_{i,j}}; \frac{z - r_{i,j} \cdot (1-\Delta_r^L)}{\Delta_r^L \cdot r_{i,j}}\right)\right) \ \forall i,j \tag{7.19}$$

the membership function of the fuzzy set $\tilde{r}_{i,j}$ that is the actual ratio between product i and j; let also be

$$\neg\xi_{\tilde{r}_{i,j}} = \min\left(1; \max\left(\frac{r_{i,j} - z}{\Delta_r^L \cdot r_{i,j}}; \frac{z - r_{i,j}}{\Delta_r^R \cdot r_{i,j}}\right)\right) \quad \forall i, j \tag{7.20}$$

the membership function of the opposite of $\tilde{r}_{i,j}$; finally, let be

$$\mu_{r_{i,j}} = \min\left(1; \max\left(\frac{z - (r_{i,j}^* - \delta_{r_{i,j}})}{\delta_{r_{i,j}}}; 0\right)\right) \quad \forall i, j \tag{7.21}$$

the membership function of the ratio requirement of volumes i and j. Therefore the satisfaction of the requirement (7.21) given the actual ratio of (7.19) computed by (7.1) and (7.2) is respectively:

$$\pi_{r_{i,j}}(\tilde{r}_{i,j}) = \pi(\mu_{r_{i,j}}/\xi_{\tilde{r}_{i,j}}) \;\; \forall i, j \tag{7.22}$$

$$\eta_{r_{i,j}}(\tilde{r}_{i,j}) = \eta(\mu_{r_{i,j}}/\xi_{\tilde{r}_{i,j}}) \;\; \forall i, j \tag{7.22bis}$$

and, as the reader can notice, it depends on the decision variable $\tilde{V}_i$ and $\tilde{V}_j$. In all the formulae (7.7) through (7.22) z is the generic independent variable that allows the mathematical formulation of both actual fuzzy sets and requirements.

7.4.2 PRODUCTION

Production deals with the definition of the manufacturing system characteristics, i.e. technological capacity, production capacity and configuration.

While technological capacity basically depends on the typology of the items effectively produced in the manufacturing system, i.e. $\Im = \sum_{i \in I} x_i$, production capacity and configuration are strictly depending on the system efficiency.

When we refer to system efficiency here, we actual mean design efficiency that is different from production one. Production efficiency depends on bottleneck existence, while design efficiency depends on planned resources utilisation and therefore on the system configuration. At this design phase we do not know how products will be effectively processed within the manufacturing system, therefore it is not possible to locate bottlenecks that is an essential step to analyse production efficiency. On the contrary, based on expert knowledge is possible to link resources utilisation, i.e. design system efficiency, and system configuration.

System efficiency requirement

After having stressed this concept, let us indicate with W the set of different technological operations necessary to manufacture the whole set I of the products. If each technological operation is performed on a different workstation we say that the manufacturing system has a *dedicated resources configuration*. In this case each product has a routing corresponding to the number of different technological operations required and as a result the products throughput time is the summation of the technological, transportation and waiting times. The ratio between the summation of the technological times and the throughput times, called *Flow Index*, is widely assumed as a parameter for evaluating the system efficiency. In fact, the more the product flow within the manufacturing system is complicated, the lower is the system efficiency. In a dedicated resource configuration this index is generally very unsatisfactory.

If, on the other and, all the technological operations are grouped into flexible workstations, a *general-purpose resources configuration* is obtained. The products have shorter routing tables and as result the waiting and transportation times are reduced, causing an increment of the flow index value. Therefore, moving from a pure dedicated configuration to a *pure general-purpose* one, the efficiency of a manufacturing system, measured by the flow index, increases.

At this stage of the design it is almost impossible to predict product throughput times therefore, the flow index measure cannot be used to evaluate the efficiency of the IMS. It needs to be replaced by another index that is an efficiency measure estimable at the actual design stage. Let us indicate with *Flow Complexity Index* (FCI_i), the summation of the routing stages of each product i. The value of FCI_i depends on the configuration of the manufacturing system, so that, if k is the number of the possible alternative configurations, each product will be characterised by a FCI_i^k. Of course, the above index depends on which product is effectively manufactured in the system. Since, in our simultaneously approach, we are interested in considering the flow of the products effectively running in the system, i.e. $\Im$, the following *configuration flow complexity index* is defined:

$$FCI^k(x_i) = \sum_{i \in \Im} FCI_i^k \tag{7.23}$$

$FCI^k(x_i)$ is directly correlable to the flow index and in the same time is easily computable at the design stage, therefore it can be assumed as a measure of the system efficiency at this stage of the design. Furthermore, as the reader can notice, it depends on the decision variables x_i.

A basic requirement production aims to obtain can be stated as the *improvement of the system efficiency as much as possible.* This requirement allows to reduce throughput times, WIP, delivery delays and so forth. Because of the discussed connection between system efficiency and production flow, the above requirement can be also obtained by *simplifying the production flow as much as possible.* Let us indicate with $FCI^{\min}$ and $FCI^{\max}$ respectively the minimum and the maximum value of $FCI^k(x_i)$, i.e.:

$$FCI^{\min} = \min_k(FCI^k(x_i)) \tag{7.24}$$

$$FCI^{\max} = \max_k(FCI^k(x_i)) \tag{7.25}$$

By using the above indexes, the vague requirement "*improving the efficiency of the manufacturing system as much as possible*", can be formalised as depicted in Figure 7.5. For each configuration k, the degree of satisfaction of the requirement in Figure 7.5 can be formalised as it follows:

$$\pi_{prod}(y^k, x_i) = y^k \cdot \max\left(0; \min\left(\frac{FCI^{\max} - FCI^k(x_i)}{FCI^{\max} - FCI^{\min}}; 1\right)\right) \tag{7.26}$$

where y^k is the configuration decision variable that is equal to 1 if the configuration k is chosen, 0 otherwise. Also in this case, being $FCI^k(x_i)$ the summation of dummy

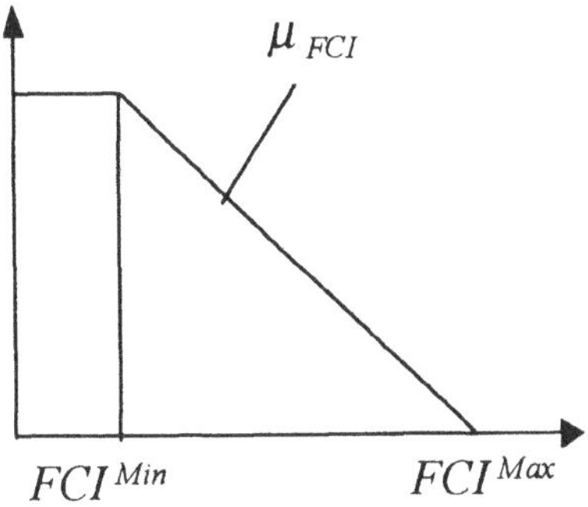

FIGURE 7.5. System efficiency requirement

variables, possibility and necessity modalities give the same result. It is straightforward that the satisfaction of requirement (7.26) leads towards general purpose configurations.

System efficiency versus production capacity

Moreover we also need to characterise the dependence of the production capacity from the configuration through the system efficiency. This characterisation is necessary to evaluate the investment.

The dependence of the resources capacity from the system efficiency is very well known. In fact, given the total workload that each technological resource has to trigger in a fixed production horizon, the number of resources, for that technological resource, is generally obtained by dividing the workload times the resource availability in the planned horizon. Not considering here reliability problems, the above approach implicitly assumes that each resource is available for production for its entire availability. This is quite false, since effective resources utilisation depends on the system design efficiency and therefore configuration. It is well known, indeed, that the more a manufacturing system is efficient the more the resources utilisation tends to one, meaning that each resource tends to be utilised for its entire availability. When we refer to production efficiency, resources utilisation indexes depend on the ratio between the resources workload and the bottleneck one, and consequently they are different each other. On the other hand, when we refer to design efficiency, resources utilisation indexes are just depending on the system configuration and therefore they can be assumed equals for all the resources. Hence we assume a unique system *Efficiency Index* (EI) varying in the range $[0,1]$. Under this hypothesis, the resource number is computed dividing the workload by the index EI.

The index EI is related to the system complexity, i.e. $FCI^k(x_i)$. The mathematical law linking EI and $FCI^k(x_i)$ is vaguely known. In fact, only imprecise and qualitative relationship can be used to express such a linkage. The qualitative relation between the two indexes can be easily expressed by using fuzzy linguistic reasoning. In this case, indeed, a set of fuzzy linguistic rules is used to codify such

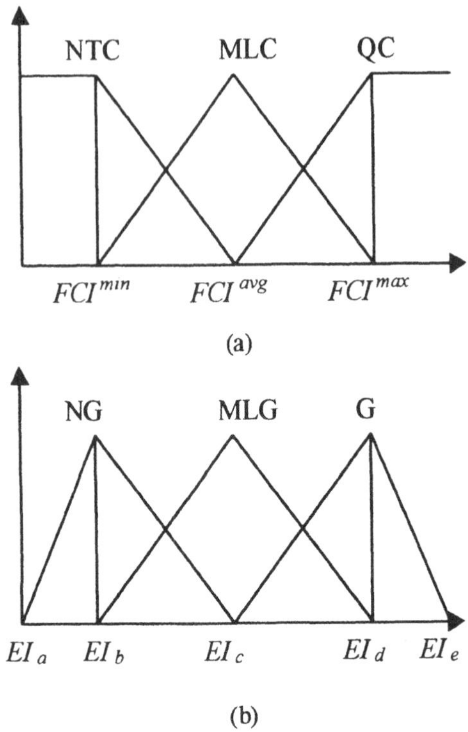

FIGURE 7.6. $LS_{(FCI)}$ and $LS_{(EI)}$

imprecise knowledge. The codified knowledge can be obtained either from expert evaluation or from a learning process. How this knowledge can be codified is not the aim of this research. What it is important here is that, being fuzzy sets universal approximators [19], a set of fuzzy linguistic rules can codify the relation existing between EI and FCI^k. Let us indicate, then, with $LS_{(FCI)}$, $LS_{(EI)}$ respectively the linguistic set associated to $FCI^k(x_i)$ and to EI. The membership functions of these two linguistic sets, reported in Figure 7.6a and 7.6b, are:

$LS_{(FCI)}$= *(Not That Complex, More or Less Complex, Quite Complex)= (NTC, MLC, QC)*
$LS_{(EI)}$= *(Not Good, More or Less Good, Good) = (NG, MLG, G)*

and where:
$FCI^{avg} = \left(FCI^{(\max)} + FCI^{(\min)}\right)/2.$

Using the above linguistic sets, the qualitative relations among $FCI^k(x_i)$ and EI, can be expressed by the following set of fuzzy rule:

Fuzzy Relation between System Complexity and Efficiency

S1: if FCI is NTC then EI is G
S2: if FCI is MLC then EI is MLG
S3: if FCI is QC then EI is NG

Now we can use Mamdani et al. [24], inference approach to figure out a value of EI for a given value of $FCI^k(x_i)$. Let us suppose to fix an input value for FCI (the reader should refer to Figure 7.7). This value actives the antecedent of the above rules in different ways $(A1, A2, A3)$. Those values cut the fuzzy sets, $F1, F2, F3$, on the rule consequence as depicted in Figure 7.7. The union of the above fuzzy sets produces a composed fuzzy set. The effective EI value, corresponding to the actual $FCI^k(x_i)$ basing on the above linguistic set, is given computing the Centre of the Area (COA) of the composed fuzzy set. Let us indicate this value as $EI(FCI^k(x_i))$.

The resulting value of $EI(FCI^k(x_i))$ is, now, used to compute the effective number of workstations to be considered in the manufacturing system. Let us indicate with WL_w^k, the workload resulting on a generic workstation w, when the configuration k is chosen. This value is given by the following expression:

$$WL_w^k = y^k \cdot \sum_{i \in \Im} a_{i,w}^k \cdot t_{i,w}^k \cdot V_i \tag{7.27}$$

where:
$a_{i,w}^k$, is 1 if the product i has a stage on the workstation w when the configuration k is chosen, 0 otherwise;
$t_{i,w}^k$, is the machining time of i on w, when the configuration k is chosen;
V_i, is the production volume required for the product i.
By the above considerations the effective "*design workload*" is given by:

$$DWL_w^k = \frac{WL_w^k}{EI(FCI^k(x_i))} \tag{7.28}$$

Indicating with A the resources availability in the production horizon T, the number of workstations w in the configuration k to be considered is given by:

$$N_w^k(x_i, y^k, V_i) = Int\left[\frac{DWL_w^k}{A}\right] \tag{7.29}$$

Eq. (7.29) clearly states that N_w^k depends on the decision variables x_i, y^k, V_i. It should be noticed that V_i could be a fuzzy variable. In this case a fuzzy multiplication operator should be used in (7.27) leading to a $\tilde{N}_w^k$, i.e. a fuzzy value of N_w^k.

Relation (7.28) shows how production efficiency influences the determination of the resource number and therefore, the amount of the investment.

7.4.3 Redditivity and Risk

Redditivity

Redditivity deals with the investment capacity of generating profits. Since several measures can be considered to address investment redditivity, as first thing the

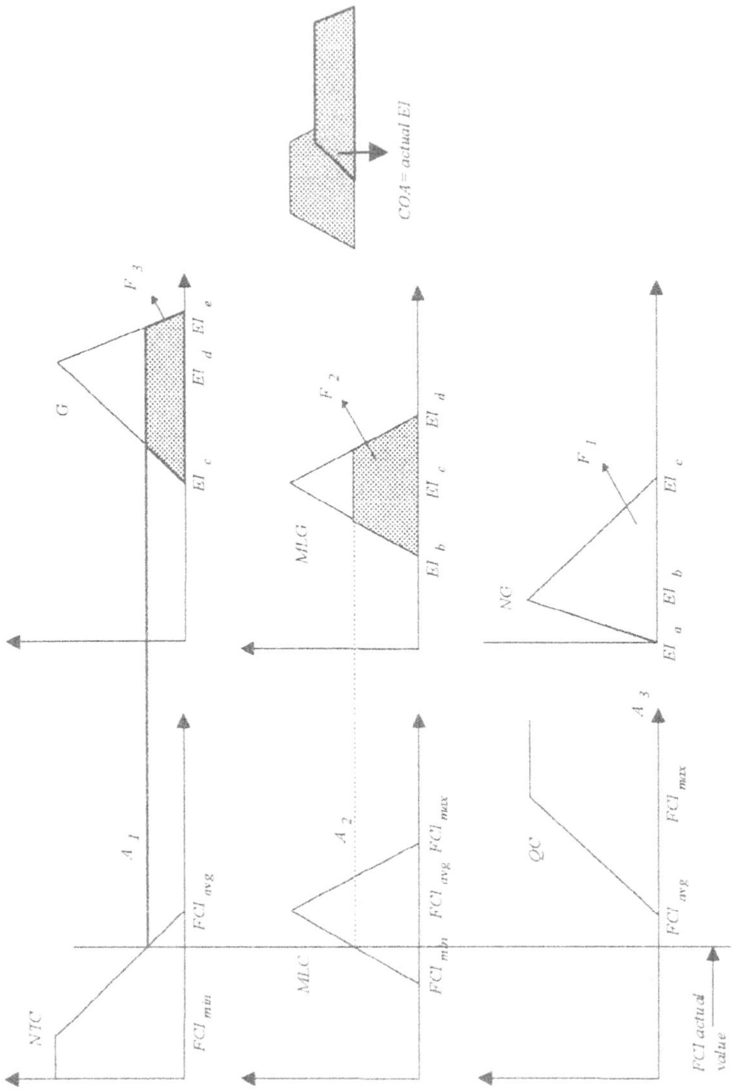

FIGURE 7.7. System Complexity versus Efficiency Index Fuzzy Inference

management is required to choose the redditivity measure. Let us consider, as redditivity measure (RM), the ratio between the net present value of the investment gains (NPV) and the Amount of the Investment (AoI).

The investment amount is easily computable knowing the resource investment cost when the configuration k is chosen, C_w^k. Therefore, the following expression gives the value of AoI:

$$AoI^k = \sum_w N_w^k \cdot C_w^k \tag{7.30}$$

The NPV can be easily computed by the following expressions:

$$NPV^k = \sum_{t=1}^{n} \sum_{i \in \Im} \frac{(p_i - c_i^k) \cdot V_i}{(1+\alpha)^t} \tag{7.31}$$

where:
n, is the investment life (years);
α, interest rate to be applied;
c_i^k, is the unit cost of the product i when the configuration k is chosen.

Finally, for each configuration k, the Redditivity Measure is given by:

$$RM^k(x_i, y^k, V_i, p_i) = \begin{bmatrix} \frac{NPV^k}{AoI^k} \ if \ y^k = 1 \\ 0 \ otherwise \end{bmatrix} \tag{7.32}$$

Eq. (7.32) underlines the dependence of RM^k from the decision variables. It is important to notice that c_i^k depends on the actual value of N_w^k and therefore, through (7.27)-(7.29), from the configuration k; so that:

$$c_i^k = c_i^k(y^k) \tag{7.33}$$

The relation (7.33) depends on the accounting method used to compute the industrial costs. On the above formulae several items can be approximate known and hence fuzzy, for example V_i, p_i and c_i^k, but overall n and α. This last problem is one of the most discussed in the literature of advanced manufacturing systems economic justification [16, 21, 31]. The decision-maker, indeed, really does not have a precise evaluation both for n and for the value of the interest rate, α, along the investment life. Hence, those values are really fuzzy and they can be formalised using triangular fuzzy numbers with a left and right spread equal to a fixed percentage, $\Delta\%$, of the central value. If such spread is not big, the values of NPV^k and c_i^k can be still considered triangular fuzzy numbers [43] even if power, multiplication and division operations are not conservative of the linear shape. Under this hypothesis RM^k is still a triangular fuzzy number, even if not necessarily a symmetric one, hence it can be denoted with the usual notation $\tilde{RM}^k = (RM_a^k, RM_b^k, RM_c^k)$.

Let now be the redditivity satisfaction function defined how depicted in Figure 7.8. In order to figure out the degree of satisfaction of the requirement given the actual investment redditivity,$\tilde{RM}^k$ let us indicate with:

$$\xi_{\tilde{RM}^k} = \max\left(0; \min\left(\frac{RM_c^k - z}{RM_c^k - RM_b^k}; \frac{z - RM_a^k}{RM_b^k - RM_a^k}\right)\right) \quad \forall k \tag{7.34}$$

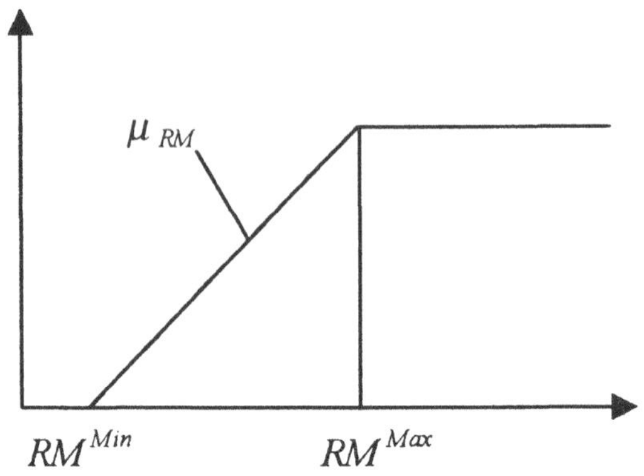

FIGURE 7.8. Redditivity satisfaction function

the membership function of the fuzzy set $\tilde{RM}^k$ that is the actual investment redditivity; let also be

$$\neg\xi_{\tilde{RM}^k} = \min\left(1; \max\left(\frac{RM_b^k - z}{RM_b^k - RM_a^k}; \frac{z - RM_b^k}{RM_c^k - RM_a^k}\right)\right) \quad \forall k \tag{7.35}$$

the membership function of the opposite of $\tilde{RM}^k$; finally, let be

$$\mu_{RM} = \max\left(0; \min\left(\frac{z - RM^{\min}}{RM^{\max} - RM^{\max}}; 1\right)\right) \tag{7.36}$$

the membership function of the redditivity requirement. Therefore the satisfaction of the requirement (7.36) given the actual redditivity of (7.34) computed by (7.1) and (7.2) is respectively:

$$\pi_{RM}(x_i, y^k, V_i, p_i) = \pi(\mu_{RM}/\xi_{\tilde{RM}^k}) \ \forall k \tag{7.37}$$

$$\eta_{RM}(x_i, y^k, V_i, p_i) = \eta(\mu_{RM}/\xi_{\tilde{RM}^k}) \ \forall k \tag{7.37bis}$$

Risk

Risk usually deals with the time a project requires to recover the fix investment. Therefore, as a measure of the investment risk, can be assumed the Payback Period (PP) of the investment computed as the number of year, PP, needed to recover AoI. Also PP , basically for the same reasons of RM, is generally a not symmetric triangular fuzzy number $\tilde{PP}^k = (PP_a^k, PP_b^k, PP_c^k)$ depending on the chosen configuration.

Let now be the risk satisfaction function defined as depicted in Figure 7.9. In order to figure out the degree of satisfaction of the requirement given the actual

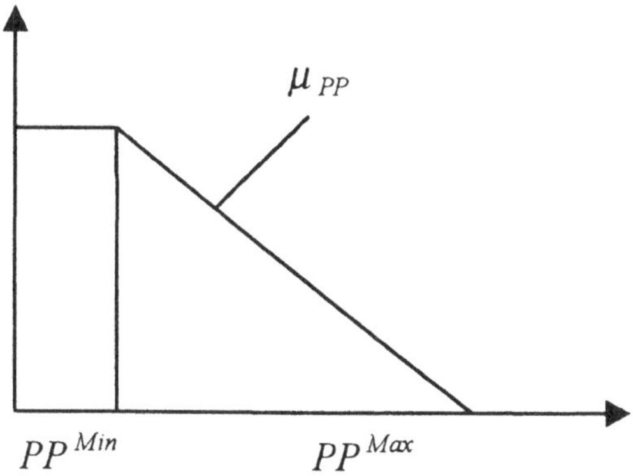

FIGURE 7.9. Risk satisfaction function

investment risk measured by $\tilde{PP}^k$ let us indicate with:

$$\xi_{\tilde{PP}^k} = \max\left(0; \min\left(\frac{PP_c^k - z}{PP_c^k - PP_b^k}; \frac{z - PP_a^k}{PP_b^k - PP_a^k}\right)\right) \quad \forall k \tag{7.38}$$

the membership function of the fuzzy set $\tilde{PP}^k$ that is the actual investment risk; let also be

$$\neg\xi_{\tilde{PP}^k} = \min\left(1; \max\left(\frac{PP_b^k - z}{PP_b^k - PP_a^k}; \frac{z - PP_b^k}{PP_c^k - PP_a^k}\right)\right) \quad \forall k \tag{7.39}$$

the membership function of the opposite of $\tilde{PP}^k$; finally, let be

$$\mu_{PP} = \max\left(0; \min\left(\frac{PP^{\max} - z}{PP^{\max} - PP^{\max}}; 1\right)\right) \tag{7.40}$$

the membership function of the risk requirement. Therefore the satisfaction of the requirement (7.40) given the actual risk of (7.38) computed by (7.1) and (7.2) is respectively:

$$\pi_{PP}(x_i, y^k, V_i, p_i) = \pi(\mu_{PP}/\xi_{\tilde{PP}^k}) \ \forall k \tag{7.41}$$

$$\eta_{PP}(x_i, y^k, V_i, p_i) = \eta(\mu_{PP}/\xi_{\tilde{PP}^k}) \ \forall k. \tag{7.41bis}$$

7.4.4 Framework optimisation model

Based on the assumed hypotheses the framework for the strategic IMS design can be summarised as it follows:

Decision variables: $x_i, y^k, V_i, p_i, N_w^k$
Goals:
1. Maximise marketing strategy requirement satisfaction: max[(7.7)];
2. Maximise production volume requirement satisfaction: max[(7.11) *or* (7.11*bis*)];

3. Maximise production price requirement satisfaction: max[(7.15) *or* (7.15*bis*)];
4. Maximise production mix requirement satisfaction: max[(7.22) *or* (7.22*bis*)];
5. Maximise production efficiency requirement satisfaction: max[(7.26)];
6. Maximise redditivity requirement satisfaction: max[(7.37) *or* (7.37*bis*)];
7. Maximise risk requirement satisfaction: max[(7.41) *or* (7.41*bis*)];

The choice among which expression to use for each objective, i.e. if normal or *bis* equations, depends on the decision-maker risk propensity versus the uncertainty. Optimistic decision-makers will prefer possibility measure, therefore normal expressions, while pessimistic ones will choose necessity and hence *bis* expressions.

7.5 Numerical example

In order to make clearer the proposed approach and to show its actual applicability to real problem a very simple numerical example has been carried out. Referring to the proposed framework described in the previous sections the following assumption has been assumed:

- The firm needs to design a IMS able to manufacture up to 6 different products. Products 1 and 2 have to be manufactured in any case.

- The marketing strategy of the firm has been summarised such as: "*the company is fully satisfied when at least four products are manufactured and it is completely unsatisfied when only two products are made*". This requirement has been formalised as depicted in Figure 7.4a where $I^* = 4$ and $\delta_I = 2$.
- Product volume strategy does not admit any vagueness, therefore the production volumes are fixed and crisp, and their values have been reported in Table 7.2.
- Product price strategy does not admit any vagueness, therefore the product prices are fixed and crisp, and their values are reported in Table 7.2.
- There are no mix requirements.
- The products can be manufactured in five different basic workstations. If this configuration is adopted the routing and the machining times matrices will be those reported in Table 7.3. However, the operations performed by the workstations 1, 2, 3 can be grouped together into flexible machine as well as the operations performed by the workstation 4 e 5. Combining all of the cases coming out from the grouping possibilities, 10 different design configurations have been obtained. Those configuration possibilities have been reported in Table 7.4. Each one of these configurations has its routing table. These new tables are obtained from the one reported in Table 7.3, by associating the operations grouped. Service times of the new configuration tables have been obtained by considering that, when two operations are grouped together the resulting machining time is the summation of the single service times reduced by 0.5, in order to consider the saved set-up time.

Product	Volumes ($\cdot$ 100)	Price (CU)
1	90	18.0
2	140	16.0
3	80	8.0
4	170	8.5
5	110	20.0
6	100	28.0

TABLE 7.2. Volumes and prices

- No system efficiency optimisation requirement, see (7.26), has been considered.
- Concerning the relation between system complexity and the efficiency index, the linguistic variables and values of EI and FCI are those depicted in Figure 7.6a and 7.6b where the basic points of EI in Figure 7.6b are respectively 0.5, 0.65, 0.75, 0.85, 0.9. The fuzzy rule set is the one discussed in paragraph 7.2.
- Investment costs of each basic manufacturing resources have been assumed equal to 225,000 Currency Units (CU). When two operations are grouped together in a flexible manufacturing resource, its investment cost is obtained by the summation of the two resource costs reduced by a factor (assumed equal to 10% of the total investment cost) in order to take account for scope economies.
- Product unit costs have been computed considering only the manufacturing cost by using the following expression that particularise the (7.33) for our example:

$$c_i^k = \sum_w c_w^k \cdot N_w^k \cdot t_{i,w}^k \tag{7.42}$$

  where c_w^k is the cost per unit of time of the resource w when the configuration k is chosen. Such cost has been obtained by:

$$c_w^k = \frac{1}{A} \cdot \left[C_w^k \cdot \frac{(1+\alpha)^n \cdot \alpha}{(1+\alpha)^n - 1} \right] \tag{7.43}$$

  where $\alpha = 0.1$, $n = 10$ years and $A = 230,400$ units of time and C_w^k have been previously discussed. Being N_w^k a function of DWL_w^k through (7.29), c_i^k depends on the actual value of EI.
- The values of $RM^{\min}$ and $RM^{\max}$ have been assumed equal respectively to 0.5 and 1.0.
- The values of $PP^{\min}$ and $PP^{\max}$ have been assumed equal respectively to 2 and 5.
- Satisfaction requirements have been computed by using necessity measure and therefore, the *bis* expressions.

Product	WS1	WS2	WS3	WS4	WS5	WS1	WS2	WS3	WS4	WS5
1	1	1	1	0	0	7	5	2	0	0
2	0	1	1	0	0	0	4	8	0	0
3	0	0	0	1	1	0	0	0	5	2
4	0	0	0	1	0	0	0	0	9	0
5	1	1	0	0	1	3	6	0	0	7
6	1	0	1	1	1	8	0	3	3	6

TABLE 7.3. Routing and service times

Configuration	WS1	WS2	WS3	WS4	WS5
1	1	2	3	4	5
2	1-2	3	4	5	-
3	1	2-3	4	5	-
4	1-3	2	4	5	-
5	1-2-3	4	5	-	-
6	1	2	3	4-5	-
7	1-2	3	4-5	-	-
8	1	2-3	4-5	-	-
9	1-3	2	4-5	-	-
10	1-2-3	4-5	-	-	-

TABLE 7.4. System possible configurations

Concluding the decision problem has fourteen dummy decision variables, four product variables in order to decide which product has to be considered, i.e. x_i, and 10 configuration variables, y^k. The problem has been modelled as explained in section 5.4. It has been optimised by using a very simple genetic algorithm. The result for this problem is reported in Table 7.5.

Even if very simple, the numerical example highlights the spirit of the proposed approach. As the reader can notice the solution of the model has selected products 3 and 6. The resulting configuration is the number six. It is characterised by a flexible resource, where either operation 4 or 5 can be performed. This because products 3

Products	*Proposed approach*	*Classical approach*(1)	*Classical approach*(2)
Products	1,2,3,6	1,2,5,6	1,2,3,4
Configuration	6	7	1
Number of resources	1,1,1,1	2,1,1	1,1,1,2,1
Value of EI	0.7	0.76	0.63
Marketing strategy satisfaction	1	1	1
Redditivity satisfaction	1	1	1
Risk satisfaction	1	0.76	0.33

TABLE 7.5. Results of the numerical example

and 6 have a very similar routing table, especially referring to last operations and moreover, because the summation of their production volumes, together with a good Efficiency Index (0.7), allows obtaining a workload on the flexible workstation 4-5 that guaranties a good resource utilisation. Therefore, combining operations 4 and 5 in a flexible machine both scale and scope economies are gained and, in this way, good redditivity and risk coverage performances are obtained.

The above result has been compared with a classical hierarchical approach in which the firm first selects products and afterwards designs the optimal manufacturing system for those products. Let us suppose that the company, based on market considerations, selects products 5 and 6 (see Table 7.5 classical approach (1)). The optimal configuration in this case is the seventh. Such configuration, having two flexible resources, allows to gain a better value for EI (0.76). But, as the reader can notice, the satisfaction of the risk requirement is not fully gained. This is because the aggregate production volume of products 5 and 6 does not result enough high to produce a workload that justifies, in term of risk, two flexible workstations.

Let us suppose now that, products 3 and 4 are chosen (see Table 7.5 classical approach (2)). In this case the products does not fit very well with other products in the systems and therefore, the best configuration that is possible to locate is the first one. This, being the configuration with only dedicated resources has a very low EI and therefore leads to a low satisfaction of financial and risk requirements.

Even in this very simple example, the location of an optimal solution is not a trivial problem and the example shows how only a simultaneous approach to the problem can lead to a full satisfaction of all the requirements.

7.6 Conclusions

This research proposes a new approach for Integrated Manufacturing Systems design. The essential issues characterising the newness of the proposed approach can be summarised as it follows:

- *a simultaneous approach in front of a classical hierarchical one.* Usually, IMS design is performed postponing technological and production decisions to market and financials ones. Since IMS investments consist in high fixed costs and long life cycles, the classical approach can lead to a poor manufacturing efficiency and, therefore, to a low long term redditivity. We propose a simultaneous approach in which marketing, production and redditivity and risk requirements participate all together to the investment decision;

- *decision environment uncertainty formalisation.* In our approach the uncertainty characterising the decision environment is not brought back to a crisp formalisation that does not allow to consider its actual complexity and vagueness. We propose possibility formalisation to model imprecise requirements and unknown data and relations;

- *a quantitative decision tool.* The use of possibility theory allows us to turn imprecision and vagueness in a programming model whose optimisation provides solutions to effectively made design decisions.

The simple numerical example developed shows how the proposed approach can be effectively implied as a decision-making tool in IMS decisions.

Acknowledgements
This research has been founded by MURST.

7.7 References

[1] Antonsson, E. K., and Otto, K. N.., 1995., Imprecision in Engineering Design. *ASME Journal of Mechanical Design.* Vol. 117.

[2] Burstein M. C.. 1986. Finding the Economical Mix of Rigid and Flexible Automation for Manufacturing Systems. *Proceedings of the 2nd ORSA/TIMS Conference on Integrated Manufacturing Systems: Operation Research Models and Applications.* K. E. Stecke and R. Suri (eds.), Elsevier Science Publisher, Amsterdam.

[3] Buzacott J. A. and Mandelbaum M.. 1983. Flexibility and Productivity in Manufacturing systems. *Proceedings of the annual IIE Conference*, Los Angeles CA.

[4] Buzacott J. and Yao D.. 1986. Integrated Manufacturing Systems: A Review of Analytical Models. *Management Science*, Vol. 32, No. 7.

[5] Carrie A.. 1988. *Simulation of Manufacturing Systems.* John Wiley and Sons Ltd.

[6] Dallery Y. and Frein Y.. 1986. An Efficient Method to Determine the Optimal Configuration of a Integrated Manufacturing System. *Proceedings of the 2nd ORSA/TIMS Conference on Integrated Manufacturing Systems: Operation Research Models and Applications* , K. E. Stecke and R. Suri (eds.), Elsevier Science Publisher, Amsterdam.

[7] Davis, E.. 1987. Constraint propagation with interval labels. *Artificial Intelligence*, Vol. 32.

[8] Dong, W. M. and Wong, F. S.. 1987. Fuzzy weighted averages and implementation of the extension principle. *Fuzzy Sets and Systems*, Vol. 24, No. 2.

[9] Dubois D. and Prade H.. 1979. Fuzzy real algebra: Some results. *Fuzzy Set and Systems*, Vol. 2.

[10] Dubois, D.. 1987. An Application of Fuzzy Arithmetic to the Optimization of Industrial Machining Processes. *Mathematical Modelling*, Vol. 9.

[11] Dubois, D., and Prade H.. 1991. Fuzzy set in approximate reasoning, Part 1: Inference with possibility distribution. *Fuzzy Sets and Systems*, Vol. 40.

[12] Finch, W.W. and Ward, A.C.. 1995. Generalized set-propagation operations over relations of more than three parameters. *Artificial Intelligence in Engineering Design, Analysis, and Manufacturing*, Vol. 9.

[13] Garret, S. E.. 1986. Strategy first: A case in IMS justification.*Proceedings of the 2nd ORSA/TIMS Conference on Integrated Manufacturing Systems: Operation Research Models and Applications*, K. E. Stecke and R. Suri (eds.), Elsevier Science Publisher, Amsterdam.

[14] Giachetti, R. E., Young, R. E., Roggatz, A., Eversheim, W., Perrone, G.. 1997. A Methodology for Reduction of Imprecision in the Engineering Design Process. *European Journal of Operational Research*, Vol 100.

[15] Goldhar J. and Jelenik M.. 1983. Plan for Economies of Scope. *Harvard Business Review*, Vol. 61, No. 6.

[16] Hundy B. B. and Hamblin D. J.. 1988. Risk and assessment of investment in new technology. *Int. Jour. of Prod. Res.*, Vol. 26, N. 11.

[17] Inuiguchi M. and Ichihashi H.. 1990. Relative modalities and their use in possibilistic linear programming. *Fuzzy Set and Systems*, Vol. 35.

[18] Kim, K., Cormier, D., O'Grady, P. and Young, R.. 1995. A System for Design and Concurrent Engineering Under Imprecision. *Journal of Intelligent Manufacturing*, Vol. 6.

[19] Kosko, B.. 1996. Additive fuzzy systems: from function approximation to learning. *Fuzzy Logic and Neural Network Handbook*, Edited by C. H. Chen, McGraw-Hill.

[20] Kotha S. and Orne D.. 1989. Generic Manufacturing Strategies: A conceptual synthesis. *Strategic Management Journal*, Vol. 10.

[21] Krinsky I. and Miltenburg J.. 1991. Alternate method for the justification of advanced manufacturing technologies. *Int. Jour. of Prod. Res.*, Vol. 29, N. 5.

[22] Lee E. S. and Li R. J.. 1993. Fuzzy multiple Objective Programming and Compromise Solution with Pareto Optimum. *Fuzzy Set and Systems*, Vol. 53.

[23] Lo Nigro G., Noto La Diega S., Perrone G.. 1995. Modelli per la Progettazione Strategica dei Sistemi Flessibili di Lavorazione. *Procedure di Gestione delle Risorse nei Sistemi Integrati di Produzione, Ed. P.E. Corti and Klinger - Collana MATMIX*, Padova.

[24] Mamdani E. H. and Efstathiou H. J.., 1986. *Expert Systems and Optimisation in Process Control.* Unicorn Series, Technical Press. Aldershot.

[25] Michewicz, Z..1996. *Genetic Algorithms + Data Structures = Evolution Programs.* Third Springer, New York, New York.

[26] Naik B. and Chakravarty A. K.. 1992. Strategic acquisition of new manufacturing technology: a review and research framework. *Int. Jour. of Prod. Res.*, Vol. 30, No. 7.

[27] Navinchandra, D., and Rinderle, J.. 1990. Interval Approaches for Concurrent Evaluation of Design Constraints. *Proceedings of Concurrent Product and Process Design, San Francisco* , CA, ASME Publication DE-21.

[28] Nelson C. A. and Knight L. B.. 1986. A Mathematical Programming Formulation of Elements of Manufacturing Strategy: IMS Applications. *Proceedings of the 2nd ORSA/TIMS Conference on Integrated Manufacturing Systems: Operation Research Models and Applications* , K. E. Stecke and R. Suri (eds.), Elsevier Science Publisher, Amsterdam.

[29] Otto, K. N., and Antonsson, E. K. . 1993. The method of Imprecision Compared to Utility Theory for Design Selection Problems. *DE-Vol. 53, Proceedings of the fifth Int. Conference on Design Theory and Methodology*, ASME, Sept. 19-22, Albuquerque, NM.

[30] Pendlebury A. J.. 1987. Creating a Manufacturing Strategy to suit your Business. *Long Range Planning*, Vol. 20, No. 6.

[31] Perrone G. 1994. Fuzzy Multiple Criteria Decision Model for the Evaluation of AMS . *CIMS Journal*, Vol. 7.

[32] Perrone G. and Noto La Diega S. 1996. Strategic IMS Design Under Uncertainty: A Fuzzy Set Theory Based Approach. *Int. J. of Prod. Econ.*, Vol.46-47.

[33] Perrone, G. and Young R.E.. 1997. The use of Fuzzy Possibilistic Programming in dealing imprecision in the early phases of product design *Proceeding of AITEM III*, Fusciano (SA).

[34] Porter M. E.. 1987. *Competitive Strategy*, The Free Press - New York.

[35] Reusch, B.. 1993. Potentiale der Fuzzy-Technologie in Nordhein-Westfalen. *Studie der Fuzzy-Initiative NRW*, Ministerium fr Wirtschaft, Mittelstand und Technologie des Landes Nordhein-Westfalen, Dsseldorf.

[36] Skinner W.. 1985. *Manufacturing: The formidable Competitive Weapon.* John Wiley and sons, New York.

[37] Solberg J.. 1978. Analytical Performance Evaluation for the Design of Integrated Manufacturing Systems. *Proceedings of the IEEE Conference on Decision and Control*, San Diego CA.

[38] Solberg J. and Nof S.. 1980. Analysis of Flow Control in Alternative Manufacturing Configurations. *Journal of Dynamic Systems, Measurement and Control*, Vol. 102, No. 141.

[39] Stam A. and Kuula M.. 1991. Selecting a Integrated Manufacturing System using Multiple Criteria Analysis. *Int. J. of Prod. Res.*, Vol. 29, No. 4.

[40] Suri R. and Hildebrant R.. 1984. Modelling Integrated Manufacturing Systems using Mean Flow Analysis. *Journal of Manufacturing Systems*, Vol. 3, No. 1.

[41] Swamidass P. M.. 1986. Manufacturing Flexibility: Strategic issues. *Discussion Paper 305*, Graduate School of Business, Indiana University, IN.

[42] Swamidass P. M. and Newell W. T.. 1987. Manufacturing Strategy, Environmental Uncertainty and Performance: a Path Analytical Model. *Management Science*, Vol. 33, No. 4.

[43] Terceno A., Marquez N., Barbera M. G.. 1994. Funcion de pertenencia del termino amortizativo de un prestamo a interes incierto. *Comunication Papers of SIGEF I*, Vol. I, Reus.

[44] Trappey, J.-F., Liu, C. R., Chang, T.-C.. 1988. Fuzzy non-linear programming: Theory and application in manufacturing. *Int. J. of Prod. Res.*, Vol. 26.

[45] Wood, K. L., Otto K. N., Antonsson, E. K.. 1992. Engineering design calculations with fuzzy parameters. *Fuzzy Sets and Systems*, Vol. 52.

[46] Young, R. E., Perrone, G., Eversheim, W., Roggatz, A.. 1995. Fuzzy Constraint Satisfaction for Simultaneous Engineering. *Annals of the German Academic Society for Production Engineering*, Vol. II, Issue 2.

8

Adaptive Production Control in Modern Industries

Kenneth N. McKay[1]
John A. Buzacott[2]

ABSTRACT
Production planning and control is a combination of three factors: organizational elements, mechanics of plan generation, and plan execution. In rapidly changing situations, special demands are placed on each of these facets. For example, it can be difficult to establish reliable estimates for processing times when the situation constantly changes. Different production control methods and concepts are needed when the manufacturing system is faced by internal uncertainty compared to situations that are reasonably stable and predictable. As a plant races through its adolescent stage, or undergoes other self-inflicted changes, existing production control methods must also be changed to match the internal uncertainty. We discuss the underlying assumptions supporting traditional production control and use insights from the history of production management to develop a strategy for adaptive control practices.

8.1 Introduction

You have just acquired a factory, perhaps several factories. The factories might be new or existing facilities. You want to control the production process to ensure that the right items are made at the right time in the right quantities and are of the right quality; and you want to do this in the most effective and efficient fashion. What to do and how to do it? The answer will depend on a number of factors, three of which are: i) your definition of production control, ii) your type of industry, and iii) the current state of the industrial process. Production control encompasses the detailed scheduling and dispatching of work, the planning of work release, various inventory policies, and longer term aggregate planning. It also entails the organizational design of the production control function, wage and incentive policies that impact productivity, performance metrics used to measure and gauge production, and aspects of facility design that can impact productivity. These facets are interrelated and do not exist in isolation from one another - they must be balanced and complement each other for long term production control to be truly effective and efficient.

An obvious key factor for success in production control is the matching of meth-

[1]Faculty of Business Administration, Memorial University of Newfoundland, St. John's, NF, Canada, A1B 3X5, kenmckay@plato.ucs.mun.ca
[2]Faculty of Administrative Studies, York University, North York, ONT, Canada, M3J 1P3, jbuzacot@bus.yorku.ca

ods and techniques to the manufacturing situation. There have been and continue to be process, repetitive, and job shop industries. In Operations Management, a great deal of work has been done on the repetitive job shop, assembly lines, and process industries. The importance of categorizing the type of production is not a new idea. Babcock (1917) identified seven types of production systems requiring different production control methods which could be found in various combinations in almost any industry. He noted that companies had to have different control methods for each situation if manufacturing efficiency was to be realized for each of the seven major categories (pp. 126-127):

1. One order for one piece. The piece never to be reproduced.
2. One order for several pieces, never to be reproduced.
3. Repeat orders at irregular intervals for one or few pieces.
4. Repeat orders at irregular intervals for many pieces.
5. Repeat orders at uniform intervals for one or a few pieces.
6. Repeat orders at uniform intervals for many pieces.
7. Continuous or standing orders for the same piece.

Babcock was discussing the application of Taylor's Scientific Management (Taylor 1911) at Franklin Manufacturing and explained how the concepts were adapted for each situation encountered at the plant. For example, lot sizing was not the same for each category: by 1912 Franklin had some parts which were high volume, repetitive, and required at uniform intervals. Workers at Franklin determined production lot sizes for these parts by taking information such as frequency of setups into consideration (pp. 125). It is as important today as it was in the early part of this century to understand the situation in order to apply or adapt the appropriate methods. In this paper, we will discuss various forms of production and suggest two concepts that extend the classification schema and philosophy. First, we propose that major portions of industry today have an element of inherent and pervasive uncertainty that is not accommodated by traditional production control models, approaches, or categories. Second, because of the uncertainty, otherwise stable and predictable situations exhibit the characteristics normally associated with the early stages of learning curves. We suggest that because of this uncertainty and instability, adaptive control practices must be developed that are sensitive and responsive to the causes and effects of this uncertainty. An understanding must also be developed as to when and how to apply the adaptive strategies. The following sections will describe the foundations of traditional production control and then will develop concepts for how to view modern production control - encompassing organizational issues, facility design interrelationships, plan generation, and execution.

8.2 Motivation - Inherent Uncertainty

Most of today's factories are faced with some form of external uncertainty (e.g., supply and demand, environmental influences) and various capacity planning and inventory control policies have been developed in response (e.g., Song and Zipkin 1996). There can also be uncertainty within the plant (e.g., Daniels and Kouvelis 1992, Wein and Chevalier 1992). Process industries are typically considered to be relatively free from internal uncertainty. Hence there have been many sophisticated modeling techniques developed for the continuous process problem. Repetitive manufacturing plants may or may not be immune to internal uncertainty. Repetitive high volume plants usually become stable after a period of time. The frequency and magnitude of process and/or product changes will then dictate long term production stability. The majority of production management advice and research has focussed on process industries and mass production that is relatively stable. If this is the situation, there are many well known and proven approaches to controlling production. An unstable repetitive situation is hoped to be a short term phenomenon on its way to becoming stable mass production and is often managed with inventory in the interim. For a period of time, there would be limited uncertainty focussed on certain processes and jobs. The unique or one-off job shop is considered the hardest one to manage. It is almost impossible to use rigid plans, schedules, policies, and procedures in this environment because there is substantial and sustained uncertainty throughout the process. Two questions are raised by the above:

- What production control techniques are appropriate while a specific industry or plant is maturing and has internal uncertainty?
- What techniques are appropriate after maturity as the situation evolves and changes? Possibly changing from one category to another?

Answers to these two questions can possibly assist with two other questions:

- What techniques are appropriate for the repetitive manufacturing category if uncertainty turns out to be a long term condition?
- What techniques are appropriate for intermittent manufacturing?

In this paper, we will explore these four questions using insights derived from the turbulent years of early industrial development (early 1900's) and the relatively stable mid-century period (e.g., 1945-65 and 1965-80). These periods provide us clues as to how production control has evolved into the methods of the 1980-90's and how the challenges of production control have been addressed in the past.

8.3 Applying Production Control Methods - A Perspective

Any factory that is shipping product is using some form of production control, but it may not be appropriate, effective, or efficient. It may be ad hoc and not have even been consciously decided upon, but it is still production control. It has been our experience that many people in the production control role are unsure about what production control philosophies to use, why certain techniques are being presently used, why certain techniques give good results and why other techniques do not. They understand neither the origins of the techniques they are using nor the necessary and sufficient conditions for the techniques to function. Furthermore, they do not know how to interpret production control fads of the moment. If they are lucky, the techniques being used or adopted are appropriate and there is little harm in not knowing the why's and wherefore's. But then again, the techniques might not be the best ones to use - there can be two or more criteria in conflict and driving at cross-purposes. In the following sections, we will briefly review how production control methods were applied in the early part of this century 1900-30, the middle years 1945-65, 1965-80, and more recently 1980 to the present.

8.3.1 Production Control - 1900-1930

This period of production control was clearly dominated by the first step of any methodological approach - trying to eliminate the confusion and get a good handle on what is happening where. The literature of this era is dominated by organizational design, rate and incentive programs, facility design, the bookkeeping methods needed to control an industry, and mechanical aids to help control/track production (e.g., Kimball 1925, Diemer 1935, Emerson 1913, Gantt 1919, Jones and Hammond 1918). As noted by Emerson (1913): "... most of the industrial plants of the world are still in the stage of civilization of which as to transportation the old freight wagons and prairie schooners across the plains were types. They started when they got ready, they arrived some time, and nobody knew where they were nor what route they were taking in between." (p. 251). This general state of chaos was reflected in the writings - they had to learn how to walk before they could run. There were also a few authors talking about reasonably advanced concepts of economic lot sizing (Babcock 1917, Erlenkotter 1989, Green 1915), the elimination of waste throughout the system (Knoeppel 1911), the concept of pulling work through a factory (Knoeppel 1917), treating the next department within the plant as a customer and achieving continuous improvement in quality (Gantt 1910), and mechanical scheduling similar to Kanban concepts (Coes 1928). There were also authors talking about the importance of quality for the consumer (Hook 1928) and how to control quality (Robertson 1928a,b). However, as noted in a retrospective by Alford (1922), few firms were observed to be using basic manufacturing management techniques - we can surmise that fewer were using the advanced concepts. Two texts at the time focussed specifically on the economic analysis of manufacturing - Kimball (1929) and Raymond (1931).

By 1933, Alford (1933) was able to report that positive trends could be seen on many fronts including concepts such as industries providing service to the community in which they found themselves, work design, cost accounting, waste elimination, wage payment schemes, marketing, job standardization, material handling, and so forth. He mentioned that the two major trends of the previous decade were towards "more management per man" and specialization of the work force. A major research area noted by Alford in 1933 was the use of economical analysis - "mathematical formulas ... establishment of rates and optimal levels" (p. 10). Alford stated that the formulations would have to be made simpler before general application of the principles would be seen in practice. The topics and researchers that Alford listed were: minimum cost point (Kimball), break-even point (Knoeppel, Rautenstrauch), economies of small tools (Roe), economic manufacturing lot sizes (Coes, Hagemann,

Kurtz, Norton, Raymond, Shepard, Vorlander), economic life of equipment, economic purchase quantities (Davis), economic control of quality of manufactured product (Shewhart).

Clearly, most of the early recommended practices were oriented towards the new phenomenon of high volume, repetitive, and standardized mass production. This concentration, which began about 1880, was further influenced and inspired by the phenomenal success and publicity associated with Henry Ford's assembly line at Highland Park (Ford Motor Company 1917). There appears to be very little written about intermittent job shops, customized production, and how to manage technological change in the workplace. An early article by Sweet (1885) advised management not to be surprised by the unexpected when dealing with new inventions and incorporate these uncertainties into their planning, but these ideas did not make it into the mainstream literature. Although they were aware of the problems with new processes and products, the researchers and practitioners were concentrating on how to organize the workplace and understand the process and steps required to produce a standard product (Alford 1922, 1933).

Most companies were in a mess (Emerson 1913) and sorting out the mess was not seen as a quick process - Taylor commented that it would take five to seven years to introduce scientific management in a typical factory (Taylor 1910). It is interesting that the mass production situations attracted the leading industrial engineers of the day and that many of their assumptions and conclusions are similar to those used today. For example, they assumed that standardization would lead to stability and with the relatively low amounts of internal uncertainty, the production process would be deterministic. They concluded that such a stable situation could be planned in an hierarchical fashion with little or no problem solving at the lowest levels - coordination would ensure that the workers would be provided with the right resources, the right materials, and quality work would be done at the right time (e.g. Coburn circa 1919, Gantt 1910). Therefore, they did not spend much effort on the period of time leading up to a stable situation; it was a necessary evil to deal with but not one for which a firm constructed long term organization elements or production control methods.

If one considers the types of products and processes in industry 1900-1930, it seems that the early industrial engineers were probably right to assume that most mass production processes, products, and plants would enjoy relatively long periods of stability - at least several years worth between any significant change in materials, processes, or product design. If one looks beyond the research topics, the early practitioners left behind the centralized planning department, hierarchical control, master planners, material requirement planning systems, dispatch and control systems, and tracking systems as documented in various texts (e.g., Diemer 1935 and Kimball 1925), and just-in-time material arrival (Nash 1928) to name just a few concepts. Many of the concepts evolved as the factories discovered how to deploy the methods - what was easy to do, what was not. For example, many tracking and inventory systems were proposed and experimented with. However, all of those that would give timely (i.e., daily) information for planning purposes relied on extensive manual effort and discipline on the part of the workers and management. It appears that very few firms committed to the expense or were able to exhibit the required discipline.

Then as now, widespread adoption of the concepts seemed to be dictated by the situation in which a particular industry found itself. For example, the concept of just-in-time material arrival was common from 1921 through 1927 (Dutton 1928) as firms had to survive an inventory crisis (Chandler 1977). The practice seemed to decrease with time and all but disappear; this in parallel with changes in the marketplace. Similarly, the focus on waste reduction was huge during the late 1920's with firms in one city competing with firms in another to see who could be the most efficient (Conrad et al 1930). Alas, this attention to waste also disappeared, only to resurface many decades later.

Unfortunately, many of the manual systems for production control required discipline and attention to detail, a skill North American firms were not renown for (Emerson 1913). In fact, Emerson suggested that Americans turn to Japan and Germany to learn efficiency. These early systems also required extensive amounts of labor so that results and information could be used in a timely fashion - there was a paper explosion. Whatever the reasons, it seemed to become normal to rely little on time coordination, and use larger buckets of time with less detailed control/tracking, to use intermediate stores to decouple the various stages of production, and to push work through the factory. By using these techniques, the individual functional departments did not constrain each other. The master planners and centralized planning departments would divide up the work and let the foremen perform the dispatching and sequencing of work at the detailed level. The foremen would bargain with each other and their work force using a variety of incentive programs (e.g., Gantt 1910) to achieve their goals.

From the beginning, firms have been struggling with the establishment of performance metrics and how the operations will respond. Emerson (1913) points out the problems when the metrics are not thought through or when several metrics are in conflict with each other - work will be created and managed so that the desired numbers look good while ignoring the larger picture (p. 110). In 1910, Gantt described how he had implemented a performance scheme in one factory where one department's performance was based on how well its customers performed (i.e., the next department within the factory). Gantt described how communication between the two departments increased and how "The result is a continuous improvement in the quality of their work." (p. 159). Gantt also described how incentives had to be tied to the bigger picture and how teams and individuals could be encouraged to help each other. Unfortunately, the enlightened concepts were not widely adopted and most wage/rate schemes focussed on individual piece rates (e.g., see descriptions in Alford and Bangs 1949). Similarly, departments were viewed as independent entities which caused each area to manage their own metrics to the disadvantage of other departments whenever possible.

8.3.2 Production Control - 1945-1965

For many reasons, the situation during 1945-1965 did not look that much different from the early period or the period in between. The same types of firms existed - high/low volume, stock/custom, large/small (Hempel 1950). Plossl and Wight (1967) in their preface and introduction describe the 1930's and 1940's as a period that did not encourage scientific management since survival was the paramount concern and then business was so good after World War II that inventory control and other practices were not important for survival. In the 1930's, most plants were operating below capacity and were lacking customers, not materials or resources. This state meant that most firms during this period were able to offer short delivery times and were not faced by large inventory and control problems. In the 1940's, throughput was a high priority, as was time to market. Interest rates were quite low, so financing inventory was not a problem and inventory control policies were not focussed upon. Plossl and Wight note that it became common place for companies to chase stock via expediters - clearly a reactive approach rather than a proactive one. They proceed to comment on the general state of manufacturing: "Probably the biggest problem in applying scientific techniques in industry has been the fact that companies were not ready for these techniques because they had not even begun to solve many of their basic problems in controlling manufacturing." (p. 5). In preparing the third edition of their text, Moore and Jablonski (1969) note that many firms (some large firms, almost all of the small ones) were doing things the same way when they wrote their first edition in 1951: "Perhaps this should be expected since the procedures for making mature products in mature industries don't change much in any short span of years." (preface). Almost all the larger firms were using computers for bookkeeping and inventory control, but there was very little evidence or mention of mathematical practice in their text. Computerization in small to medium sized firms was quite rare.

The focus of practitioner and researcher alike remained on the mass production industries where the work force was progressively becoming more specialized and the organizations even more structured. The mass production firms were also large enough to hire consultants, to do research, to try out pilot projects, to get involved with computerization, and to participate in the electronics and aerospace programs - much like it is today. The large firms were also able to maintain an interest in concepts such as statistical sampling (e.g., American Statistical Association 1950). The hierarchical control structures developed in the 1920's were largely accepted as is with some discussion (e.g., Simon 1946) but without much formal analysis of their workings until Anthony's seminal work in 1965 on hierarchical production planning. There was an amazing growth period with reasonably cheap money during which North American firms were easily able to continue the past practices of intermediate stores, raw inventory stores, and inattention to waste. As before, the basic technologies, processes, and products were relatively stable and Anthony described a situation without much internal uncertainty that required little in the way of adapting or problem solving at the operational level.

Generally, computers were scarce in manufacturing and while the concepts of economic lot sizing and statistical control appeared in production handbooks (Alford and Bangs 1949), the concepts did not appear to be widely applied (Plossl and Wight 1967). In addition, the early practices of using simple dispatching heuristics, (e.g., due dates and priorities) continued and there was no detailed explanations of how to use mathematics to derive dispatch decisions in the Alford handbook or early texts of the era (Moore and Jablonski 1969). The application of mathematical analysis to production control as a research topic continued in the 1950's and 1960's with such seminal works as Holt et al (1960), Conway et al (1967), Arrow et al (1958), and Bowman and Fetter (1961). Totten (1967) presents a collection of research papers from this era that dealt with the application of quantitative techniques to facility planning. As a general statement, it appears that the academic research results were ahead of their time and there was no widespread adoption of any scientific method. For example, Plossl and Wight describe the use of linear programming for planning and scheduling and computer simulations for using dispatch heuristics, but these are discussed as advanced concepts for future exploitation.

While the practice was not widespread, the use of computers for manufacturing was steadily increasing throughout the 1950's and 1960's. Diebold (1952) talked about existing examples of automated retrieval systems, a computer being sold for ($15,000 US) that simulated a factory schedule for ten machines, the anticipated linking of suppliers and customers by electronic means, and the future of production control systems as the key aspects could become quantified. Diebold noted the challenges facing the automated factory and the need for redesigning the process and product for automated manufacturing. As computers became available and more popular, the traditional paper systems of bills of material, bill explosion, order generation, and inventory tracking were replaced. Once this information was in a form suitable for information processing, then the concepts of EOQ for inventory analysis became more common. For example, one of the authors remembers working in a domestic appliance factory in the UK in the early 1960's. The first manufacturing uses for the IBM Ramac 305 in the period 1962-64 were for level by level coding, bill of materials explosion (not time phased), inventory transaction recording, and then EOQ's for the purchased parts.

It is possible that one reason (of possibly many) that the research results were not paid their due, was because there was a perception that they were not needed - production was working. Production worked because things were inherently stable, money was cheap for decoupling inventories, the level of control matched the level that could be executed, and the hierarchical model was manageable. If it was not broken, why fix it?

8.3.3 Production Control - 1965-1980

This was the era of MRP and MRP-II. In one sense, there were great changes with significant innovations in the areas of automation, computerized information systems, material handling systems, and so forth. In another sense, little changed with respect to organizational design, performance metrics, production control methodologies, and day-to-day operational practices. The TQM, JIT, etc. programs were in the future and there were few threats perceived on the horizon.

The closed loop information systems (e.g. MRP) integrated many of the manufacturing processes, such as shipping, production, engineering, purchasing, and receiving (Wight 1974). However, the basic concepts of having a bill of material, blowing it out, netting various forms of receipts and requirements had been around in various forms since the early 1900's. If anything, computerization made it more feasible for more firms to practice the basic concepts than before.

Unfortunately, the widespread use of economical analysis for the various management functions still did not occur with the exception of EOQ type options imbedded in the MRP systems. There were significant achievements in certain industries where there were few large machines to control (e.g., steel - Lefkowitz et al 1975) but the bread and butter mass producers were still using basic machine loading and intermediate stores to coordinate production. Interactive Gantt chart editors for scheduling did not exist and there was little, if any, reported adoption of the batch oriented scheduling tools created.

During this time, a certain level of internal uncertainty existed and grew - there were significant changes in processes, materials, and products. Certainly, any firm involved with electronics, plastics, alloys, or chemicals saw dramatic changes, as did any firm involved with military projects. The high levels of inventory masked this uncertainty and cushioned the plants - the traditional production control systems worked and any costs were passed through to the customer. At a macro level, the firms would look stable and be running smoothly, but on a daily level, there might be lots of events not going as planned - inventory levels were high throughout the system and there were long lead times.

One of the reasons for this state of internal chaos was that good production control implies the availability of reliable, timely, and high quality information from the shop floor. It was not until the 1980's that bar coding was widespread and PC-based shop floor tracking systems were feasible. During the 1965-80's, the common tracking technique was to have expediters walk the floor and manually count and track what was where. This created many difficulties when accurate and timely data was needed. With time cards and manual reporting, the work force also found many ways to maximize their own incentive rewards. Work would be reported when the timing was in the worker's favour and not always when the work was performed. Unfortunately, this meant that the system data on progress bore no relation to the actual situation. Without reliable information about the internal state, all the companies could do was focus on the supply chain - upstream supply and downstream distribution.

In summary, the period 1900-1980 was dominated by relatively long periods of process and product stability with many mature industries. The mass production industries grew in size and the organizational structures and basic production control methods matched the problem situation. When there were prolonged periods of instability (e.g., the 1921 inventory crisis noted by Chandler 1977 pp. 456-457) old ways were set aside and alternative techniques were used; techniques that required a certain discipline and effort to sustain. As soon as the situation recovered, the efficiency methods were dropped and inventory became the predominant tool - if in doubt, build more. Perhaps it was easier to manage inventory than people. The pervasive use of intermediate stores also permitted the hierarchical planning model to function and to assign planners as needed to major areas with each protected by buckets and inventory buffers. It was a balanced situation and firms thrived. Although it was a high cost solution, it was still cheaper than what many foreign competitors could offer: importing raw or semi-processed materials, performing the work, and then shipping the finished products to market. As everyone now knows, this situation was not to last.

8.3.4 Production Control - 1980-present

It took several decades for firms in the Far East to get everything figured out, but they did. Once again, North American firms were again advised to look to the Japanese for lessons in efficiency, effectiveness, and discipline. The Japanese firms applied (or re-invented) many of the common sense techniques that had been written about in the early 1900's. As noted in the 1900-1930 subsection, the concept of pulling work through a plant, using mechanical scheduling techniques, and involving the workers in continuous improvement had already been preached. It is also well known how the Japanese adopted the statistical control concepts once they were introduced after World War II (Hopper 1982). Not only did North American firms scurry about trying to copy the new Japanese methods, but a wholesale attack on inventory commenced and continues to this day. This has resulted in an imbalance - the cushion has been removed and the instability has surfaced but the organizational structures and production control concepts have not really changed, remaining anchored in the mature and stable past. The concepts and approaches used by most firms have not adapted to a changing situation; the software systems do not address it, neither do management practices. They continue to use hierarchical control paradigms designed for stable mass production and have not learned how to manage sustained inherent uncertainty. Anthony (1988) recognized this phenomenon and noted the need to adapt the production system when inherent uncertainty exists. Most modern firms do not experience several years of peace and quiet any more. They are lucky if a day or week goes by without a change somewhere in the plant. If the market is not forcing a change, internal continuous improvement programs are. Any change to the status quo must be managed and while some impacts cannot be predicted, some can be (McKay 1992).

There have been changes made to the way manufacturing has been conducted, but the changes have been largely mechanistic or uni-dimensional in nature. Few of them challenge the underlying assumptions of maturity and eventual stability. For example, there have been many concepts and proposed solutions for curing manufacturing's problems, some of which are: 6-Sigma, Zero Defects, Continuous Flow, ISO 9000, Re-Engineering, BRP, BPR, ERP, MRP-II, CRP, CQI, CPI, EDI, FSS, Leitstands, Agility, Vertical Integration, Matrix Management, Teams, Best of Class, Green Field, Benchmarking, Virtual (pick a topic), Focussed Factories, FMS, Group Technology, Cells, SCM, JIT, Kanban, Pull, TQM, QSM, Single Minute Die Exchange, Kaizen, Continuous Improvement, ad nauseam. None of these concepts and programs provide a decision making structure (organization, control, and execution) for inherent uncertainty for sustained periods of time even though some of them encourage it. They all address some aspect of the problem or a symptom of a deeper problem and in that regard, have produced results. For each of these phrases, there are examples of successes, failures and there are equal numbers of prophets and naysayers. For example, Ramberg (1994) noted that Senge's 1993 keynote speech at the annual ASQC conference described that out of 500 firms surveyed, two-thirds of the TQM programs had ground to a halt. Duimering et al (1993) also note how difficult it is to sustain themes of the month and suggest how the organization should be changed before technical or procedural changes are introduced.

Too often it seems that practitioners assume that one technique or approach is the major cause for any success or failure without understanding the complete system and the possible interrelationships between the humans and technology. The search for a Holy Grail of manufacturing is not a new endeavor: Emerson (1913) described how foolish it was to think that solutions to manufacturing problems could be discovered through brief plant visits and how multiple methods and concepts would have to be adapted based on the situation. Some of the changes which have been introduced in recent years are very much on the reactive side with emphasis placed on being able to move from a weekly net requirements bucket to a day, or how to perform reactive scheduling (e.g., Bean 1991). The adaptive heuristics and models we have (e.g., Gershwin 1989) are adaptive to various output or input parameters and the heuristics themselves do not adapt - they are single stage control models and are insensitive to changes in the underlying situation.

There has also been a re-emphasis on teams and team decision making during the past decade (Duimering et al 1993). In some cases it is a convenient way for management to disclaim personal responsibility for decision making. In others it is a way to dramatically change the authority structure in an indirect approach, and in yet others, it is a way to improve communications, manage change, respect and make use of employee knowledge, improve quality, and make the organization more effective and efficient. As with other fads, teams have worked well in some situations (by design or by accident) and have failed where the concept was not properly understood and the company was not ready for the adoption.

8.4 Production Control Concepts For Immaturity Or Uncertainty

The previous section describes the underlying assumptions and requirements of current production control techniques. During this century, these techniques have proven themselves to be effective in many situations and have influenced the organizational structure of firms and the way decisions are arrived at. The inherent assumptions about problem stability also facilitated the development and refinement of many mathematical models and algorithms for exploring the problem and deriving optimal or near-optimal results. There are many situations today that still satisfy the underlying assumptions and requirements and it is in these situations that the matching solutions should be advocated and encouraged. However, there are also many situations where the assumptions are not valid and the problem is significantly different. For these cases, we must explore new approaches or apply old approaches differently. We need to adapt or evolve existing models to the changing problem or derive new adaptive models that can handle sustained and pervasive uncertainty. A number of concepts for organizational design and adaptive production control are described in the following subsections. The concepts are exploratory and preliminary and are intended to highlight the issues and possible solutions when the problem is viewed differently.

8.4.1 Organizational Design

Organizational design addresses who does what, what parts of the business an individual employee has interaction with, how authority is distributed, how information will flow through the company, and how personnel will move through the company. Consider what a typical production control situation looks like for a mature situation.

- First, a hierarchy of planners and schedulers is likely to exist. There are master planners, master schedulers, and departmental schedulers, each with well defined and developed tasks. Depending on the size of the firm, the master planners and schedulers may be responsible for all of the demand or a portion of the demand as it relates to a major product family. The planners and schedulers may be part of a centralized planning department with its own entity or be part of another general area such as materials management. The master planner works with the long term forecasts and tries to match basic plant capacity with the demand pattern - identifying peaks and troughs. The master scheduler takes over the demand as the orders firm up and is usually responsible for the short to mid-term loading of work into major time buckets and departments. The departmental schedulers take the larger time buckets and determine feasible short term loads for the departments possibly by sequencing or by using standardized information such as production rates, man hours available, etc. The area foreman or supervisor then has the responsibility to sequence work as they see best. If it is not possible to complete all the work, the planning department provides guidance as to priorities and action plans. The rationale is that the supervisor knows best how to utilize the resources. The objective of the planners is one of giving enough work to a department to keep it busy.

- Second, the degree and type of interactions with other parts of the firm are well structured. There are routine reports, meetings, and chains of command that guide operational tasks. For example, the master planner works with the sales department and product management using forecasts and projections. The master planner also works with the major commodity buyers to ensure a long term supply of key materials and with industrial engineering to ensure that reasonable standards and estimates are used. The master scheduler works with the purchasing group to make sure the flow of material matches the demand. At the next level, the departmental scheduler works with maintenance, buyers, quality control, product engineers, process engineers, industrial engineers, shipping, supervisors, and operators to make sure everything will go smoothly. Often there are regular daily or weekly meetings where various subgroups of the operational team meet. While there may be some overlap between the planning levels, the levels are segregated since each higher level guides and constrains the next lower. One of the original purposes of the centralized planning department was to free the foremen and supervisors from the non-value added activities of sorting out what is coming, what to do next, what should be priority work, etc. (Taylor 1911). Prior to the central department, the foremen had to do all of the planning and control in addition to the detailed supervision and running of the shop. Obviously, the central department would cut down on the number of people running around trying to understand the demand, understand the process routings for the work, and so forth.

- Third, the authority and responsibility associated with a task is determined via the organizational design. In the production control area the authority can be constrained by time horizons (master planner working on a yearly horizon, master scheduler 3-4 months, detailed scheduler 1-3 weeks), by product family, and by departmental decomposition within the plant. The scope associated with authority limits visibility and accountability - the decision maker is given a problem to solve, not the job of identifying or generating the problem.

- Fourth, the organizational flow of command will directly create channels of information. The flows will be top to bottom, bottom up, and across peer levels. Usually, a higher level will not see the details of lower level operations but aggregated summaries and reports. If information must cross departmental boundaries, the information might have to flow up to the appropriate level of peer cross over and then down to the level requiring the specific information In a mature situation, the paths and expectations will be known.
- Fifth, through its incentive and reward structures the organizational design will influence how people progress through the ranks. The game will be established and the rules widely understood. Each individual will know what is important to the firm as a whole and for the individual's task. It is easy to talk about targets for the day, week, month, fiscal quarter, and year - what is expected and how well things are doing.

This picture has worked and continues to work in mature, repetitive situations for a number of reasons. There is a history and the status quo is clearly understood. In addition, the situation lends itself to widely held assumptions and models for how things work and will continue to work - there are relatively few unknowns. This permits people to sit in relative isolation, process facts and be secure with the knowledge that the facts and future expectations bear some resemblance to reality. There is little need for feedback and learning mechanisms. Hence, many of today's plants can be considered adolescents and not mature adults. In addition, many mature plants are introducing change at a rapid pace through programs such as continuous improvement. What style of organization appears to work in these situations? We can obtain insights from a variety of sources: project management, learning theory, and systems dynamics.

The planners and schedulers must be in a situation to see, understand, and contemplate the system. They must also be in a situation that supports, encourages, and facilitates feedback and learning. This is not the traditional environment. If the planners and schedulers exist in the traditional type of situation, how can they understand the direct impact of change, secondary impacts which will affect other parts of the system, offset impacts which will occur in the future, and most important of all, how will they know what changes are occurring? They can not and will not. It is impossible and unrealistic to think they can or will, and their decisions will reflect this. However, it is also not reasonable to expect a single planner or scheduler to perform the decision making for the complete plant. The solution is to broaden the scope of control while at the same time reducing the product or functional control. The scope of control can be extended to cover receiving and/or purchasing through shipping. This increase in scope must be offset by the scheduler dealing with less products and subcomponents. By covering the full production cycle for a selected number of related parts, the production control takes on a project management perspective.

If the project management perspective is taken, the personnel will be deployed in a focussed factory approach with end to end responsibility for families of products.

The same number of people are used, but instead of a vertical view of the operation, they will use a horizontal view. There are three key changes to the decision making implied by this perspective; i) temporal decomposition changes, ii) information flow changes, and iii) functional decomposition changes:

Use Families - The family becomes the project - all major steps under the coordination of a single person. The individuals would have the immediate, short, and possibly mid term planning responsibility - receiving to shipping. This is a change in the temporal scope of the task. In traditional styles, various planners and schedulers deal with different time units - day, week, month, several months. This temporal decomposition prevents timely feedback as to the impact of decisions; what decision was good, what decision was poor. The project approach allows the decision maker to see the results of their decisions at various levels of execution and have the opportunity to learn from the feedback.

Bridge The Decision Layers - The project style allows the decision maker to use information from the operational levels to guide decision making at the tactical. This is a simultaneous bottom-up and top-down approach to information flow, constraint generation, and constraint relaxation. The traditional hierarchical production planning frameworks are primarily top-down models and assume that guidance and knowledge comes from above (Anthony 1965, 1988). When the situation is changing and maturing, it is likely that the lowest levels will learn what works, what does not work, and what the impacts are before the higher levels understand what is happening - if they ever hear about it at all. The truth is on the shop floor. In a project style, the scheduler/planner is tightly coupled to more than one level and thus low level knowledge and feedback can be used at the higher levels of planning. This results in plans which are better in terms of feasibility, slack placement, and contingency plans.

Use A Supply Chain Perspective - The functional responsibilities and expertise of the schedulers/planners is also affected in the project approach. In the traditional model with three schedulers, one scheduler might be responsible for the press shop, one for assembly, one for steel - each having responsibility for all work in that area. By having end to end responsibility, the work would be split into three groupings with each scheduler having to plan and coordinate the press, assembly, and steel aspects for their products and product components. This requires either dedicated areas (true focussed factories) or good communication and collaboration between the schedulers. It also requires coordination for any shared components. With a scheduler/planner having end to end responsibility, they can be aware of what is in the internal supply chain and how the final assemblies are driving the primary operations. They know what is coming and what is going and can better orchestrate the plant's resources. This knowledge and control is important with the adolescent plant or one dedicated to life-long learning. Understanding the pending and current changes in one part of the supply chain is crucial if the other parts of the supply chain are to balance and adjust to the changes in a proactive fashion.

In summary, while the hierarchical decomposition of decision making has been and remains adequate for mature and stable situations, a more project oriented style appears to be needed when a firm is in the start up phase or is faced by rapid and sustained changes. This approach is feasible and easily managed when production can be changed into product-focussed cells or dedicated lines per product. If multiple schedulers have to compete for shared resources, the management task is not impossible but requires excellent communication and objectives that drive the decision making from the global system level and not at the individual product. In some situations, it might be necessary to manage the large families with a project style and dedicated resources while having some of the plant operate in the hierarchical fashion.

8.4.2 Plan Generation

Production control is an iterative process of planning and execution. Plans are generated and by themselves accomplish nothing - there is no value in a plan by itself until it is executed. A good plan will stipulate how the firm's resources can be used effectively and efficiently while satisfying anticipated demand. Since almost every decision that results in a plan involves compromises and tradeoffs, the plan's goodness depends on how good the general decision making process is. For example, a good decision making process must be capable of identifying the decisions that need to be made, understanding what the constraints are, knowing how the constraints interrelate, how the constraints can be manipulated, what the impact is when constraints are manipulated, identifying and using appropriate solution methods, and recognizing when the best solution has been reached.

As has been noted above, traditional plan generation is an isolated affair with a relatively formal and rigid structure. On the input side, there are the usual facts about what is needed when and the routing information that allows timing and sequencing to be established. Also on the input side, there is feedback from the factory floor. Plan generation typically uses this information in a closed-loop control fashion. For example, based on what happens on the factory floor, estimates are changed for processing times, projected yield, expected completion times, batch sizes, job release, and subsequent job starts. This is a single stage control loop where the feedback information is incorporated with the requirement data and resource information as input into decision algorithms. State dependent heuristics are of this form (e.g., Gershwin 1989). The single stage approach assumes that the control mechanism itself is stable and does not change.

Unfortunately, in our learning and changing situation, changes are needed in the decision process itself, as well as, in the output of that process - the decisions. The generating process must exhibit the characteristics of a two stage control model with a high level control stage controlling the decision making controller that generates the actual decisions. The control mechanism must have sensing and filtering mechanisms to obtain the necessary inputs and decision feedbacks for both levels of control. Such a two stage decision model has been proposed in McKay et al (1995).

In this model a preliminary sensing and filtering mechanism separates normal decision making information from the information that changes the decision making itself. The normal information is processed by the stage two control logic - the decision logic that is active at the time using the current objective and constraint information. The information that affects the decision logic itself is processed by the stage one control logic. The stage one logic adapts heuristics, creates new heuristics, senses risk, tunes algorithms, relaxes constraints, establishes policies, and modifies objectives. The learning and adapting is performed by the stage one controller and without it, the decision logic is the same day after day without any context sensitivity. This two stage decision framework has been observed in the field and has been shown to be adaptive and receptive to changing environments (McKay 1992). While it might appear obvious that if the subject of decisions is changing and dynamic the decision process must also be dynamic and capable of change, there are few planning models which incorporate this aspect implicitly or explicitly.

In addition to the decision process being dynamic and adaptable, the plan should also be sympathetic to a changing situation. There are certain concepts which can be imbedded into schedules and plans which acknowledge internal uncertainty:

Placement Of Slack - The intelligent placement and magnitude of slack (idle) time is one such concept. If changes are planned, idle time can be inserted in strategic places to accommodate potential impacts. The idle time will reduce schedule nervousness, minimize schedule regeneration, and improve the schedule predictiveness (O'Donovan 1997).

Batch Splitting - Another concept is intelligent batch splitting in which small batches are run prior to a full batch of work. Depending on the number and nature of changes since the part was last run, one or more small batches can be run to requalify and prove out the system. Any impact on processing time, yield, and quality will be absorbed by the small batch and not by the main job. This batch splitting concept can be associated with changes to products, processes, and resources.

Production Tests - A refinement on the batch splitting concept is analogous to fire drills and training exercises for disaster response units. At regular intervals, small batches of work that stress the system are run. For example, in many factory situations, there are types of work that push the machines and resources to the limit for tolerance, complexity, and other processing characteristics. If this work is produced periodically, it is likely that any subtle impacts will be exposed - documentation out of date, inadequate training, misadjusted gauges, etc. The added cost of running this work can be partially offset if the products can be used for satisfying some future demand, but the added cost can be justified if the products are expensive and the cost of possible miscues during production are significant.

Timing - A concept that acknowledges the impact of learning and change is the consideration of secondary and tertiary support services as constraints. For example, when generating a plan, any change should be probably scheduled for Tuesday through Thursday on the day shift. This avoids the typical Monday and Friday issues and ensures that the engineering and support staff are available during the shift.

Risk Avoidance - Avoiding the anticipated impact on key work is also possible. Critical work can be ahead or pushed out to avoid a period of time that is expected to be unsettled. The converse is also sometimes possible - scheduling changes to avoid critical work. For example, it might be possible to restrict many industrial engineering changes to one or two weeks of the month and avoid peak demand periods or critical windows.

These plan generation concepts address what is placed on the plan, when it is placed on the plan, and how much is placed on the plan. During periods of stability, these ideas are in the background and normal best or optimal strategies can be followed. When changes or change impacts are anticipated, the adaptive plan generation concepts can be activated. These can be viewed as extreme examples of state dependent heuristics but a better analogy is what is called a piggy-back heuristic (McKay et al 1995). The adaptive heuristic sits above or beside the normal heuristic and modifies the priority or decision criteria when appropriate. The piggy-back heuristic is activated by some event or sensory input (machine changeover or repair) and remains active for a period of time until the system reaches normal parameters.

8.4.3 Plan Execution

The plan is for the next twenty-four hours or the next week or for some horizon. When the factory is operating, the plan is being executed and if the factory is mature and stable, it is reasonable to expect a high degree of correspondence between what is expected and what transpires. This correlation reduces the need for extensive rescheduling strategies and similar reactive techniques.

When the factory is in our learning and changing mode, two themes are foremost - rescheduling and reconciliation. Even with all of the tricks from plan generation, there will be unexpected impacts and surprises that will have to be reacted to. Rescheduling is important because the plan's integrity is at risk and if the plan was originally good, the maximum of the remaining goodness should be retained. Maintaining the plan is also important when key material, resource, and delivery decisions have been made and are not easily modified. In these latter cases, it is not always possible to completely reschedule the world at each unexpected event. Rescheduling or schedule repair has been considered by Bean et al (1991) and Smith (1991). Rescheduling research has traditionally used the mainstream scheduling assumptions, heuristics and measurements. It is suggested that rescheduling research must also consider the issues highlighted in the plan generation section. When rescheduling because of an unexpected event, some work should be pushed away from the impact zone if at all possible (McKay et al 1995), batches may be split, and so forth. In the real world, constraints have a certain elasticity which can be exploited when reacting to the unforeseen.

Reconciliation is what happens after the plan has been executed for a while. In a stable situation, the unexpected does not happen too often and the reconciliation process is relatively short and can be largely automated. During reconciliation, the quantity expected versus produced is processed, and the start and finish times are tracked. When the approximate quantity is made on the machine where it is expected and at the time it was expected, reconciliation is indeed easy. However, if the unexpected occurs often, special allowances are required. For example, adequate time and support tools are required to pick up the necessary information and make the adjustments. Things can be so dramatically different in the morning compared to when the scheduler left at the end of the previous day that the result is almost a complete makeover. In other words, not much of the schedule happened as planned and the immediate activities are significantly different from those envisioned fifteen hours prior - different demand has to be satisfied with a different resource mix. For example, it is not that unusual to have a part made that was not on the plan or a different resource than expected do the work at a different time. It is possible to setup the machine and close out the job without making a single part. It is not unusual for more or less of a part to be made, or to have nothing made when something was expected to be made - even though the total allowed time was consumed. It is also possible that a damaged part can be turned into a different part with a little creativity (i.e., a pipe of a certain length can be shortened and be made into a saleable good with a different part code). Almost everything that is expected to be reported and tracked can be reported as something different when there is learning and the system is not 100% predictable. The flexibility to make these updates in an efficient fashion is a key requirement for scheduling tools and the tools cannot dictate what can or cannot be reported as having happened.

The feedback and adjustment of the process is also important when rates and run sizes can be adjusted in response to changes. For example, the dynamic control hierarchy of Gershwin (1989) provides different types of responses to different types and frequencies of events.

8.5 Conclusion

Modern industry is susceptible to sustained and rapid changes - either due to phenomena such as continuous improvement or the inherent nature of the industry. This changing environment carries with it potential impacts and uncertainty that traditional production tools are ill suited for. The evolution of production control during this century has been used to explain the underlying assumptions and origins of current practices. The organizational structure, plan generation approaches, and plan execution methods of the past must be changed and extended to incorporate the impact of anticipated and unexpected change. The production facilities of today resemble the early factories of yesteryear and not the mature facilities of the mid-century.

We believe that planning and scheduling in the situations we describe will be based on a mixture of techniques that reflect:

1. the technological features of the processes/products;

2. the capabilities, motivation/incentives and behavior of the workers;

3. managerial goals and values; and

4. managers perceptions and hunches about the future (jobs to come, process problems, people problems, etc.)

Existing scientific approaches to scheduling (primarily mathematical or simulation models) focus very much on (1) with a single, simple goal in (3) and a very mechanistic view of (2) and a simple probabilistic approach to allowing for (4). This is all necessary in order to use mathematical models. However, the harsh reality is that multiple approaches are needed to understand, research, and deal with planning and scheduling. For example, McKay (1992) used a combination of Cognitive Science, Organizational Behavior, Information Systems, Cybernetics, and Operations Management in order to understand the scheduler. Each approach typically makes certain simplifying or unifying assumptions about the world and then proceeds to develop a methodology or approach consistent with those assumptions. As a result, each approach will only give a partial picture of the situation - a situation that is multi-dimensional. The first challenge is to recognize the need for multiple methods of inquiry, the second to identify and use appropriate methods, the third to merge and integrate the results, and the fourth to use the insights gained to improve what happens in the real world of the firm.

As initial and potential areas for improving planning and scheduling, we have suggested that a project style organization for production control in conjunction with a number of proactive heuristics for plan generation and execution. An extensive field study (McKay 1992) has shown that such an adaptive control environment can exist and minimize the impacts of rapid change. We have specifically suggested that models that acknowledge and incorporate the concept of dynamic change are required for planning and executing work in today's environment and that the models must be both reactive and proactive. The planning and execution steps must recognize the risk associated with changes in the status quo and actively try to avoid or minimize the negative impact. Research on these adaptive concepts is required to better understand the issues and to develop prescriptive frameworks and models that can be used to effectively manage firms of today and tomorrow.

8.6 Acknowledgments

This research has been supported in part by NSERC grant OGP0121274 on Adaptive Production Control and by NSERC grant OGP0138270 on Modelling and Analysis of Manufacturing Systems. We would like to think Reha Uzsoy for his comments on an earlier draft and the referees for their constructive and critical suggestions.

8.7 REFERENCES

[1] Alford, L.P. (1922) Ten Year's Progress In Management, *Trans. of the ASME*, 44, 1243-1296.

[2] Alford, L.P. (1926) Laws of Manufacturing Management, *Trans. of the ASME*, 48, 393-438.

[3] Alford, L.P. (1933) Ten Year's Progress In Management, 1923-1932, *Trans. of the ASME*, 55, 7-21.

[4] Alford, L.P. and J.R. Bangs (Eds) (1949) *Production Handbook*, Ronald Press, New York.

[5] American Statistical Association (1950) *Acceptance Sampling - A Symposium*, ASA, Washington.

[6] Anthony, R.N. (1965) *Planning and Control Systems - A Framework For Analysis*, Harvard Business School Press, Boston.

[7] Anthony, R.N. (1988) *The Management Control Function*, Harvard Business School Press, Boston.

[8] Arrow, K.A., Karlin, S., and Scarf, H. (eds) (1958) *Studies in the mathematical theory of inventory and production*, Stanford Univ. Press.

[9] Babcock, G.D. (1917) *The Taylor System In Franklin Management*, The Engineering Magazine, New York.

[10] Bean, J.C., Birge, J.R., Mittenthal, J., and C.E. Noon (1991) Matchup Scheduling With Multiple Resources, Release Dates and Disruptions,*Oper. Res.*, 39, 470-483.

[11] Bowman, E.H. and R.B. Fetter (1961) *Analysis for production management*, Irwin.

[12] Chandler Jr., A.D. (1977) *The Visible Hand - The Managerial Revolution in American Business*, The Belknap Press, Cambridge Mass.

[13] Coburn, F.G., (circa 1919), Scheduling: The Co-ordination of Effort, in *Organizing For Production and Other Papers on Management 1912-1924* (Mayer, I, ed.), 1981, Hive, Easton PA.

[14] Coes, H.V. (1928) Mechanical Scheduling, in *110 Tested Plans That Increased Factory Profits* (Dutton, H.P. ed), McGraw-Shaw, New York, 69-74.

[15] Conrad W.L., Newcomb R.E., Kelsey G.W., Hagemann C.E., Ferguson W.B. and C.E. Lytle (1930) Progress in Industrial Management, *Trans. of the ASME*, 52(Part II), 1-4.

[16] Conway R., Maxwell W. and Miller (1967) *Theory of scheduling*, Addison Wesley.

[17] Daniels, R.L., and P. Kouvelis (1995) Robust Scheduling to Hedge Against Processing Time Uncertainty in Single-stage Production, *Mgnt. Sci.*, 41(2), 363-376.

[18] Diebold, J. (1952) *Automation - The Advent Of The Automatic Factory*, Van Nostrand.

[19] Diemer, H. (1935) *Factory Organization and Administration*, 5th Edition, McGraw-Hill, New York (Reprint).

[20] Duimering, P.R., Safayeni, F. and Purdy, L., (1993) Integrated Manufacturing: Redesign the Organization before Implementing Flexible Technology, *Sloan Management Review*, 34(4), 47-56.

[21] Dutton, H.P. (ed) (1928) Are We Overdoing Inventory Control - editorial from January 1927, in *110 Tested Plans That Increased Factory Profits* (Dutton, H.P. ed), McGraw-Shaw, New York, 180-182.

[22] Emerson, H (1913) *The Twelve Principles of Efficiency*, The Engineering Magazine, New York.

[23] Erlenkotter, D. (1989) An Early Classic Misplaced: Ford W. Harris's Economic Order Quantity Model of 1915, *Mgmt. Sci.*, 35(7), 898-900.

[24] Ford Motor Company (1917) *Factory Facts From Ford.*

[25] Gantt, H.L. (1910) *Work, Wages and Profits*, The Engineering Magazine, New York.

[26] Gantt, H.L. (1919) *Organizing For Work*, Harcourt, Brace and Howe, New York.

[27] Gershwin, S.B. (1989) Hierarchical Flow Control: A Framework for Scheduling and Planning Discrete Events in Manufacturing Systems. *Proceedings of the IEEE*, 77(1), 195-208.

[28] Green, J.B. (1915) The Perpetual Inventory in Practical Stores Operation, *The Engineering Magazine*, 48(6), 879-888.

[29] Hempel, E.H. (1950) *Small Plant Management, ASME Research Study*, McGraw- Hill, New York.

[30] Holt, C.C., Modigliani ,F., Muth, J.F., and Simon, H.A. (1960) *Planning production, inventories and workforce*, Prentice-Hall.

[31] Hook, J.W. (1928) Salability, Value and Better Quality - Three Prime Aids To Larger Profits, *Factory and Industrial Management*, 76(3), 479-483.

[32] Hopper, K. (1982) Creating Japan's New Industrial Management: The Americans As Teachers, *Human Resource Management*, Summer 1982, 14-34.

[33] Jones, F.D. and E.K. Hammond (1918) *Shop Management Systems*, The Industrial Press, New York.

[34] Kimball, D.S. (1925) *Principles of Industrial Organization*, 3rd Edition, McGraw-Hill, New York.

[35] Kimball, D.S. (1929) *Industrial Economics*, McGraw-Hill, New York.

[36] Knoeppel, C.E. (1911) *Maximum Production in Machine-Shop and Foundry*, The Engineering Magazine, New York.

[37] Knoeppel, C.E. (1917) *Installing Efficiency Methods*, The Engineering Magazine, New York.

[38] Lefkowitz, I., Cheliustkin, A., and Kelley, D.H., (1975) State-of-the-Art Review of Integrated Industrial Systems Control, (I. Lefkowitz and A. Cheliustkin, Eds.), *Integrated Systems Control In The Steel Industry, Conference Proceedings, IIASA*, 13-158.

[39] McKay K.N. (1992), *Production Planning and Scheduling: A Model For Manufacturing Decisions Requiring Judgement*, PhD Dissertation, University of Waterloo, Department of Management Sciences.

[40] McKay K.N., Morton T.E., and P. Ramnath (1995), *Aversion Dynamics - Scheduling When The System Changes*, Working Paper.

[41] Moore, F.G. and R. Jablonski (1969) *Production Control*, 3rd Edition, McGraw-Hill, New York.

[42] Nash, C.W. (1928), Purchasing For A Fast Rate Of Turnover, in *110 Tested Plans That Increased Factory Profits* (Dutton, H.P. ed), McGraw-Shaw, New York, 169-173.

[43] O'Donovan, R (1997) *Predictable Scheduling and Aversion Dynamics For A Single Machine*, Master's Thesis, Department of Industrial Engineering, Purdue University.

[44] Plossl, G.W. and O.W. Wight (1967) *Production and Inventory Control - Principles and Techniques*, Prentice-Hall, New Jersey.

[45] Ramberg, J.S. (1994) TQM - Thought Revolution or Trojan Horse, *OR/MS Today*, 21(4), 18-24.

[46] Raymond, R.E. (1931) *Quantity and Economy in Manufacture*, McGraw-Hill, New York.

[47] Robertson, W.L. (1928a) Quality Control by Sampling - Part I, *Factory and Industrial Management*, 76(3), 503-505.

[48] Robertson, W.L. (1928b) Quality Control by Sampling - Part II, *Factory and Industrial Management*, 76(4), 724-726.

[49] Simon, H.A. (1946) The Proverbs of Administration, *Public Administration Review*, 6 (Winter 1946), 53-67.

[50] Smith, S.F., Keng, N., and K.G. Kempf (1991), "Exploiting Local Flexibility During Execution of Pre-Computed Schedules", in *Applications of AI in Manufacturing* (eds. D. Nau and C. Tong), MIT Press.

[51] Song J, and P.H. Zipkin (1996) Inventory Control With Information About Supply Conditions, *Mgmt. Sci.*, 42(10), 1409-1419.

[52] Sweet, J.E., (1885) The Unexpected Which Often Happens, *Trans. American Society of Mechanical Engineers*, 7, 152-163.

[53] Taylor, F.W. (1911) *The Principles of ScientificManagement*, Harper and Brothers, New York.

[54] Taylor, F.W. (1910) *Scientific Management* - 1972 reprint including 1910 testimony at rate case hearings before the Interstate Commerce Commission, Greenwood Press, Westport Connecticut.

[55] Totten, D.L. (1967) *Applications of Quantitative Techniques In Facilities Planning*, Master's Thesis, Georgia Institute of Technology.

[56] Wein, L.M., and P.B. Chevalier (1992), "A Broader View of the Job-Shop Scheduling Problem", *Mgmt. Sci.*, 38(7), 1018-1033.

[57] Wight, O.W. (1974) *Production and Inventory Management in the Computer Age*, Macmillan, Toronto.